SpringerBriefs on Pioneers in Science and Practice

Texts and Protocols

Volume 4

Series Editor

Hans Günter Brauch

For further volumes:
http://www.springer.com/series/11446
http://www.afes-press-books.de/html/SpringerBriefs_PSP_TP.htm

Rodolfo Stavenhagen

Peasants, Culture and Indigenous Peoples

Critical Issues

Springer

EL COLEGIO DE MÉXICO

Rodolfo Stavenhagen
Center for Sociological Studies
El Colegio de Mexico
México, D.F.
Mexico

ISSN 2194-3125 ISSN 2194-3133 (electronic)
ISBN 978-3-642-34152-6 ISBN 978-3-642-34153-3 (eBook)
DOI 10.1007/978-3-642-34153-3
Springer Heidelberg New York Dordrecht London

Library of Congress Control Number: 2012950370

Printed on acid-free paper

Springer is part of Springer Science+Business Media (www.springer.com)

I am pleased to dedicate this book to my numerous indigenous friends in many parts of the world who have taught me the deep sense of their struggles for human rights and justice so long denied

Books by the Same Author Published by Springer

- Rodolfo Stavenhagen: *Pioneer on Indigenous Rights*. Springer Briefs on Pioneers in Science and Practice, vol. 2 – presented by Ursula Oswald Spring (Heidelberg – Dordrecht – London – New York: Springer-Verlag, 2013).

- Rodolfo Stavenhagen: *The Emergence of Indigenous Peoples*. Springer Briefs on Pioneers in Science and Practice, vol. 3, Subseries with Texts and Protocols, vol. 1 (Heidelberg – Dordrecht – London – New York: Springer-Verlag, 2013).

The cover photograph shows Rodolfo Stavenhagen accompanied by his wife Elia speaking with a group of Aymara leaders in Bolivia in 2007. (The photo is from the author's private photo collection.)

Elia and Rodolfo Stavenhagen at an Aymara public ceremony in Bolivia in 2007. *Source* Personal photographic collection of the author

Acknowledgments

The author is exceedingly grateful to Hans Günter Brauch, the most efficient and dynamic editor of these three Springer books, he has prepared on the occasion of my eightieth birthday.

Contents

Chapter 1
Introduction

This is the third of three volumes in this series containing essays written by Rodolfo Stavenhagen over a period of 50 years. In this volume, devoted to critical issues related to the general topic of peasants, culture and indigenous peoples, the author discusses a number of propositions, as they were formulated at the time, regarding peasant societies, rural development, cultural policies, racism, social inequality, and the human rights of minorities and indigenous peoples, and he suggests alternative approaches to an understanding of new social forces that resist traditional forms of domination and hegemonic views of development policies. In the last four chapters he draws on his experience as United Nations special rapporteur on the human rights of indigenous peoples, a task he carried out during the first decade of this century.

Rodolfo Stavenhagen is professor emeritus at El Colegio de Mexico. From 2001 to 2008 he was United Nations Special Rapporteur on the Human Rights of Indigenous Peoples. His scholarly work has focused on social science approaches to rural and social development, agrarian structures and reforms, ethnic and race relations, the human rights and social movements of indigenous peoples, and cultural policies. He has also been active in the policy field, holding positions in government and in several international agencies. The three volumes in the Springer Series on Pioneers of Science and Practice (vols. 2, 3 and 4) contain some of his English-language writings ranging from essays produced in the early sixties of the twentieth century to his latest writings of 2012. For a biographical essay on Rodolfo Stavenhagen, see vol. 2 of this series.

R. Stavenhagen, *Peasants, Culture and Indigenous Peoples*,
Texts and Protocols 4, DOI: 10.1007/978-3-642-34153-3_1,

Discussing water rights in Aotearoa (New Zealand), 2005. *Source* Personal photographic collection of the author

Greeting in a Bushman household in South Africa, 2005. *Source* Personal photographic collection of the author

Chapter 2
Basic Needs, Peasants and the Strategy for Rural Development (1976)

Abstract During the 1970s a number of alternative viewpoints questioned the unchallenged hegemony of the theories that the unfettered market was the only one to bring about economic development. The 'basic needs' approach provided new thinking about economic growth and gave rise to important debates in university centers and multilateral organizations such as the International Labor Organization. The Dag Hammarskjold Foundation in Sweden provided support for international meetings between scholars from the North and from the South. One such effort led to the study of alternative development approaches to which I contributed the paper presented here on the needs and possibilities of poor peasants in underdeveloped countries. The idea of peasant development has been completely ignored by mainstream development analysts.

2.1 Introduction

At the 1974 World Food Conference in Rome the dire state of malnutrition of large sectors of the world's population was fully documented. Most of the undernourished people in the world live in the underdeveloped countries, and the great majority of them live in the countryside. It is not by chance that malnutrition and accompanying indicators of low living standards are associated to a large extent with agriculture, that poverty and under consumption of food are associated with

This chapter was first published in 1976, in: Marc Nerfin (Ed.): *Another Development: Approaches and Strategies* (Uppsala: The Dag Hammarskjöld Foundation): 40–65. The permission to reprint this text was granted on 18 July 2012 by Dr. Henning Melber (Executive Director), The Dag Hammarskjöld Foundation, Uppsala/Sweden.

R. Stavenhagen, *Peasants, Culture and Indigenous Peoples*,
Texts and Protocols 4, DOI: 10.1007/978-3-642-34153-3_2,

Table 2.1 Rural poverty in Third World countries (1969)

Region	Total population	Rural population	Rural population as percentage of total population	Rural population in poverty			
				Below US$50 per capita		Below US$75 per capita	
	(millions)	(millions)	(%)	(millions)	(%)	(millions)	(%)
Africa	360	280	78	105	38	140	50
America	250	120	48	20	17	30	25
Asia	1,080	855	79	355	42	525	61
Total	1,690	1,255	74	480	38	695	55

Source World Bank, 1975: *Rural Devel-opment* (Washington 1975), Annexes 1 and 3

the world's peasantry, whose function in life is presumably to produce food. The poorest countries in the world are those where most of the population lives off the land (Table 2.1).

The agricultural sector has a twofold problem in the underdeveloped countries: (a) the need to raise production in order to satisfy increasing demand for foodstuffs; (b) the need to raise rural incomes in order to satisfy the basic needs of the majority of the world's poor, the peasants.

If the agricultural development of the poor countries in the last few decades has taught us something, it is precisely that these two objectives are not necessarily related. Agricultural production, and particularly food production, has risen fairly steadily at a slightly higher rate than the world's population; yet the income of the poorest part of the population (the peasantry) has not increased accordingly. In fact, in some areas rural income is decreasing.[1]

The explanation for this must be sought in the nature of peasant production in the Third World countries.

2.2 Agricultural Production and Agrarian Structures

Agricultural production usually falls into two kinds:

(a) production for the market, which may take place (i) on large estates or plantations with salaried or servile labor; or (ii) on small farms based mainly on family labor;
(b) subsistence production and consumption by the peasant household.
Economic growth is associated with progress in the 'modem' agricultural sector, that is, in production for the market (whether local or international), and as cash-crop agriculture advances, so subsistence agriculture is thought to recede, and eventually to disappear.

[1] See: United Nations, 1975: *1974 Report on the World Social Situation* (New York: United Nations, doc. ST/ESA/24).

However, the development of cash-crop agriculture has not led to a generalized improvement of the incomes and living standards of the rural population. This is due to various reasons:

- Cash crops for export have displaced subsistence crops for local consumption, and while monetary incomes may have increased, food consumption has often decreased in the process.
- Price fluctuations of international commodities have often severely affected producers' incomes.
- Profits from cash-crop production have become concentrated in the hands of large estate or plantation owners, or merchants and middlemen.
- The high cost of modern inputs for cash-crop production has increased the debt burden of the small producer.
- Mechanization and other capital-intensive technology usually associated with the development of modern agriculture frequently displaces labor and creates a pool of landless workers.
- Monoculture for export, so characteristic of many underdeveloped areas, prevents the emergence of integrated mixed farming oriented towards the internal market and the satisfaction of local needs.

Both subsistence and commercial agriculture are carried out within a fairly wide range of different kinds of productive units. The potential for improving agricultural output and increasing the standard of living of the rural population is directly related, among other things, to the characteristics of these units in terms of their land-tenure arrangements, labor supply and relations of production, local credit and market structures, as well as cultural values governing the economic behavior of individuals and family groups.

There is nothing further from reality than the simplistic idea that by channeling more credit or providing a little bit of technical assistance, or supplying improved inputs, backward agriculture will respond by productivity leaps which will solve the problems of output and income of the rural poor. The feasibility of success of different kinds of incentives to the operator is closely linked to the various elements of the agrarian structure mentioned above.

Agricultural production is not an activity made up of a number of isolated elements which can be juggled at will by the planner or the specialist in rural development. Agriculture as an occupation and as a livelihood is a complex social and economic system. Perhaps in no other sector of economic activity are the relationships between the following elements as much interlinked as in agriculture. These elements are: labor; technology; natural resources; social organization; income; and living standards.

2.2.1 Labor

Labor in agriculture is generally of a non-specialized nature. That is, within a given ecological framework, the agricultural laborer usually carries out most if not all of the particular tasks of the production process himself. Productive efficiency does, however, require a high level of skill and specialized knowledge, but these are generally traditional skills and knowledge which are handed down from father to son and which are suited to a particular environment.

In traditional agriculture, the application of increasing amounts of labor is usually directly related to increased output, up to a point. The use of labor is determined seasonally, and periods of labor scarcity alternate with periods of labor abundance. Labor markets are unstable and unstructured. The definition of the labor force itself is a complex task; women, who play an important role in traditional agriculture, are usually not included in labor-force statistics. Other unpaid family and reciprocal labor (children, friends and neighbors who help out at certain times of peak activity) are not easily counted nor accounted for. Observers agree that disguised unemployment is one of the principal problems of agriculture in Third World countries, yet no satisfactory measures of disguised unemployment have been developed. Agriculture is often only one of various activities that rural labor engages in (the others being small trade, handicraft production, occasional seasonal jobs in other sectors). The availability of local labor for specific agricultural tasks at the required time is frequently subject to the pressures of these complementary or alternative activities. In many rural areas of the world, temporary labor migrations within the agricultural sector itself complicate the labor picture. The requirements and the availability of manpower at the local level are thus not only related to the size of farm units and the type of crop, but also to numerous elements within the wider social and economic structure.

2.2.2 Technology

Modern agricultural technology usually appears in inverse proportion to the use of labor. The mechanization of agricultural tasks on modem farms, while contributing to raising output and productivity, often displaces labor and increases human underemployment. Modem technology requires skills, credit, capital and technically optimal farm size. It is not surprising, therefore, that it is usually concentrated, in the Third World countries, in certain privileged areas and in the hands of the privileged social classes. Modern technology has been associated with plantations, estates or large farms. The introduction of modern technology among small farmers has only recently become of general concern. Even the new seed-fertilizer technology associated with the 'green revolution', which is being directly addressed to the small farmer in many parts of the world (Asia and Latin America particularly), contributes to the concentration of wealth and greater inequalities in the distribution of income.[2]

Too little attention has been given to the development of labor-intensive, low-capital technology for the traditional agriculturist. Yet it appears that much can be done by improving traditional practices through the application of skills rather than the acquisition of costly inputs. The diffusion of technological innovations in agriculture is one of the principal tasks of agricultural extension services. Observers are agreed on the difficulties and resistances that many of these programs encounter among small farmers in underdeveloped countries. The reason for this is that the adoption of technological innovations cannot be taken in isolation from other factors such as land tenure, social organization and cultural values. The literature on the subject provides many examples of cases where 'rational' innovations have been rejected by farmers because of one, or a combination, of these various factors, and not because of any 'irrational behavior' or an abstract 'traditionalism', which some authors purport to find among peasants.

Frequently the technological innovations being promoted by public or private national and international agencies turn out to be ill suited to the natural environment, the social structure or the cultural values of the target society. To this may be added the ignorance about local conditions of so-called technical experts, or their downright biases in favor of only one kind of technological development as well as their reluctance to experiment with new processes. When this leads, as it frequently does, to costly failures in rural development projects, then renewed attempts at the local level become so much more difficult the next time.

2.2.3 Natural Resources

Natural resources (mainly soil and water) are the essential ingredient in agricultural development. They may be present or absent to varying degrees at the local and regional levels, but they may also be under-utilized, or wasted or depleted through malpractice. These resources must not be seen as something simply 'given' by nature. Their use, non-use or misuse is the direct result of social and economic organization in historical perspective. It is perhaps not an exaggeration to say that the poverty of millions of peasants in the world today in areas where there is a 'lack of resources' is not so much due to natural processes (though these do undoubtedly play their part) as to the result of social and economic ones. The poverty of many Latin American peasants who work patches of eroded earth on rocky mountainsides is the direct outcome of the monopolization of the best lands by large estate owners. The recent famine in Bangladesh, while no doubt 'caused' directly by natural factors, is the indirect result of the secular cultivation of cotton

[2] See: UNRISD, 1975: *The Social and Economic Implications of Large-scale Introduction of New Varieties of Food Grain* (Geneva: Unite Nations).

and jute under the artificially imposed 'international division of labor' of colonial and post-colonial times. Famine in the Sahel countries during 1971–1972 did not come as a surprise to observers who long ago warned that the export-oriented agriculture of those countries, with the progressive weakening of the cultivation of subsistence crops, would contribute to the particularly dramatic effects of drought on the population.[3]

In other areas of the world, fertile top soils are being depleted through the uncontrolled felling of tropical forests or overgrazing, which are man-made phenomena linked to social structures, market forces and land-tenure systems. In dry areas, the increasing use of water for urban or industrial purposes has increased the cost of this resource for agriculture and has severely affected the poor farmers. The rapid expansion of areas grown with cash crops for export in many underdeveloped countries, as a result of government policy concerned with earning foreign exchange, or as a result of monetary incentives, has had negative consequences for the conservation of natural resources in some regions. In the scramble for monetary income or quick profit the judicious use of local resources has often been neglected.

Communities that used to be relatively self-sufficient not only in food, but also in local handicraft production, building materials, raw materials for clothing, herbs for medicinal use etc. (all based on the use of local resources), have become increasingly dependent upon the market for the satisfaction of their basic needs. They have become victims of a vicious circle in which they must generate ever higher incomes for their members in order to acquire at increasing prices industrial substitutes for what they used to produce themselves. In this process, entire populations (particularly the younger people who often spend much of their time outside the community) have lost the basic knowledge and skills which previously enabled them to use carefully and maintain the equilibrium of their local resources.

It is thus a mistake to attribute the depletion and misuse of local resources, as some authors do, exclusively to the demographic pressure on the land. While population growth has undoubtedly played a role in this process, the development of market relationships is surely the main cause of the increasing disequilibrium between population and resources at the local level.

2.2.4 Social Organization

Social organization basically involves land-tenure arrangements and various kinds of relations of production between individuals and social groups that have legal, cultural and historical rights and obligations relative to the productive use of land as a resource. Much of current thinking about agricultural development is biased

---

[3] See: Comite Information Sahel, 1974: *Qui se nourrit de la famine en Afrique?* (Paris: Maspero).

towards the experience of the market mechanism of the western industrialized countries, which is proposed as a 'model' for the underdeveloped nations to follow. If the model were indeed applicable universally, we would find Danish-style dairy farmers or United States cattle ranchers all over the Third World. Inasmuch as this is not the case, and to the extent that so many attempts at local and regional agricultural development have run into trouble, it is principally because of the constraints of social organization.

Whereas capitalism, as Marx pointed out, does indeed tend to substitute the cash nexus for all other kinds of social relationships, in the agriculture of the poor countries it has not been able to do away with them yet. Not only that, but frequently the introduction of capitalism in agriculture strengthened traditional mechanisms of oppression and exploitation of the labor force. There are many instances of social constraints on the 'free' development of productive forces in agricultures. To cite but a few examples: community or tribal control over the use of land; local systems of reciprocal services of a patron-client type (e.g. Indian Jajmani); traditional chieftainships which exact tribute in money or kind from the farmers (e.g. Maraboutism in Islamized western Africa); prestige spending for ceremonial purposes implying a redistribution of income (some parts of Africa south of the Sahara, Indian communities in Latin America); the demands of kinship groups on the monetary incomes of their members (many parts of black Africa); peonage and other kinds of labor services by peasants to landlords (Latin America) etc.

When agricultural production is immersed in webs of social relations the individual farmer or producer is not always in the best situation to increase his output or improve his own standard of living. This is why so often purely monetary incentives or apparently rational criteria (by Western standards) for improving agricultural productivity do not work.

While, on the one hand, certain kinds of social structures are no doubt obstacles to the capitalist development of agriculture, on the other hand, it is the capitalist development of agriculture itself which has become an obstacle to authentic economic and social development of millions of peasants in the Third World. Capitalist agriculture has increased social and economic inequalities among social classes on the land; it has concentrated wealth, power and income in the hands of landowners or middlemen, pushed small farmers off their land and turned them into marginalized, landless laborers, and substituted the idea of gain and profit for a few for the idea of survival for the many.

However, some types of social organization (mainly the basic structure of the local village) may become the pillar upon which a different kind of agricultural development can take place, through collective or cooperative arrangements and adequate planning at the local level. In many parts of the world experiments are taking place along these lines which are opening up new possibilities for the rural poor.

2.2.5 Income

Farm family incomes can be of three types: monetary income from the sale of farm produce; domestic consumption of farm produce; and complementary income from activities off the farm. Agricultural development projects in the underdeveloped countries are usually concerned with the first kind: they tend to improve the output of saleable farm commodities and the monetary incomes derived therefrom. But as has already been pointed out above, the expansion of cash-crop production frequently displaces the cultivation of local subsistence crops. Monetary income from the sale of cash-crops must be spent on food imported from other regions or even from abroad. Inflationary pressures are common, middlemen turn sizeable profits, the regular supply of foodstuffs is often not assured and the increase in monetary income is not necessarily an indicator of increase in wellbeing.

The insecurity inherent in agricultural production, due to the forces of nature as well as the price fluctuations of cash-crops for export, makes farming an uncertain proposition at best for millions of cultivators around the world. Even when they engage in the production of cash-crops, the regular flow of monetary income is not assured. But when the farmer is deeply involved in the monetary economy he regularly needs hard cash simply to survive. This is one of the main problems facing the poor farmers in the underdeveloped countries. In order to solve the basic problem of survival he falls increasingly into debt, he tends to use institutional credit, the purpose of which is to enable him to carry out his productive activity, for day-to-day consumption needs (and often neglects improvements on his field in the process), and he seeks additional income through wage labor or other activities.

The poor farmer, in order to make ends meet, seeks multiple sources of income in a regular pattern of alternate activities of which the cultivation of his own plot of land is only one. The rural poor are mainly concerned with obtaining regular income flows; farming on small plots of land under the circumstances of traditional or tropical commodity agriculture is not the best way to achieve this end.

The vicious circle of poverty in a monetary economy has a negative impact on subsistence agriculture also. In areas where not all of the farm produce goes to the market, peasants retain a part of their crop for domestic consumption. But frequently, particularly in humid climates, they lack the means for storing and conserving their cereals. Also, owing to accumulated debts and other needs, they must sell quickly to the local middlemen. Yet when their stores of food grain run out, they often must buy back their own grain later in the year at prices several times higher. This is a frequent occurrence.

Rural income is closely related to the problem of employment, which in turn is linked to land tenure and technology. In areas where labor is abundant, rural wages are usually well below legal minimum standards. Landless laborers or subsistence farmers on micro-plots will work at times for any wage, and will often travel long distances in order to find employment (in e.g. West Africa, the Andean highlands). Only if and when the benefits of increased agricultural productivity can be equitably distributed among the rural population in the form of higher real incomes for all social classes will the question of disguised unemployment on the land cease to be significant. But this is a question of social and economic organization of the wider society and not only of the setting of minimum wages or price supports.

2.2.6 Living Standards

Living standards are not directly related to monetary incomes. The relationship between these two variables is mediated by social organization and cultural values. It is still an open question whether the transformation of traditional subsistence agriculture into cash-crop farming for export (as has occurred in many underdeveloped countries) improves or rather worsens the living standards of the rural population. On the basis of material from many areas of the world, an argument can be made for the latter assertion.

The problem hinges, of course, upon the adequate definition of living standards. Increased consumer spending as a result of monetary incomes does not necessarily raise a family's or a community's level of wellbeing. At the level of the world's rural poor it is doubtful whether the mere increase in monetary incomes (which moreover usually accrue only to a small part of a community's population) will turn into improved standards of living without planned government intervention. The basic elements of satisfaction for the wellbeing of a rural collectivity are not provided through the economic activities of a few individuals. An adequate water supply, the building of an all-weather road, sewerage, housing, electricity, health services, an adequate provision of basic foodstuffs at reasonable prices, schooling and, of course, access to productive resources such as land, water, fertilizer and modem technology for the peasant masses, can only be made available to the majority of the population through concerted government action.

Thus, whereas the increase in monetary income can indeed be furthered through various well-known market mechanisms, the collective improvement of the rural poor can only be achieved through collective planning and action, which does not necessarily imply an increase in monetary incomes for poor rural families. On the contrary, where monetary incomes have been increased rapidly during a short time span and have tended to benefit only a privileged minority in the locality or the region, there we generally find that increasing inequality produces social disorganization, tensions and conflict which become the major obstacles to progressive social change for the benefit of the community as a whole.

The six basic elements that have just been discussed—labor, technology, resources, social organization, income and living standards—are crucial factors in the possibility of social and economic change at the local level for the great masses of the rural poor in the underdeveloped countries. Each one of these dimensions (and others which have not been included) presents itself differently in particular settings and is related to all the others in a complex set of interrelationships which constitute organic wholes or systems. These systems are the various kinds of agrarian structures that are to be found around the world. In order to assess the possibilities for economic and social change in agriculture at the local level, let us briefly summarize the different kinds of agrarian structures that are most common in the Third World countries today.

2.3 Customary or Communal Land Tenure Systems

In these the land is neither privately owned nor a marketable commodity, but rather controlled by the community, whose members may have traditional usufruct or access rights to it under certain specified conditions. It is usually associated with primitive technology, shifting cultivation, subsistence farming or small-scale family production of commercial crops. Under this system, permanent improvements on the land are unlikely. The availability of family labor is the main constraint on the expansion of agricultural operations. Demographic pressure reduces the land/man ratio and generates out-migrations and a tendency towards the transformation of communal tenure into individual ownership, a tendency sometimes supported by government policy. Communal land-tenure systems are widespread in Africa south of the Sahara, in the indigenous regions of Latin America and in some tribal areas of Asia.

2.3.1 The Small Peasant Farm

This is characteristic of areas with a high density of population. The farmer directly owns his land or else holds it under some form of lease, tenancy or sharecropping arrangement, and mainly works it with the help of family labor. The small farm may provide for subsistence but it is also integrated into the market through the sale of agricultural surpluses. It may also be wholly devoted to the production of a marketable crop. When the farm is held under a tenancy or sharecropping arrangement, then a large part of the farmer's output must be set aside to support a parasitic, dominant social class that exercises a legal or customary right to the peasant's produce. In such systems, landlords are not entrepreneurs but rentiers; their interest in agricultural innovation is slight; they tend to be absentee owners, politically conservative and basically opposed to modernization. In some Asian countries, a whole chain of intermediate tenants links the direct producer to the landowner; all of them live off the peasant's labor. Obviously, unless the land-tenure system changes, the peasant producer will hardly be able to improve his situation and will not be likely to respond to the conventional economic incentives designed to improve the performance of agriculture.

2.3.2 Large Feudal or Semi-Feudal Estates

These are the traditional hallmark of Latin American and Middle Eastern agriculture. Under this system most of the cultivable land is monopolized by a small landholding elite and the peasant population is tied to the estates under different kinds of servile labor arrangements or service tenancies. The laborers are usually allowed a plot of land for their own subsistence crops, but they are required to work on the estate for the owner's benefit under his direct supervision or that of special supervisors or administrators. Estate owners do not usually innovate, being content to draw a regular income from the labor of their attached peasant workers. Estates are generally managed quite inefficiently, and much of the land is underutilized. Technology remains traditional, and is mainly that of the peasants themselves.

Estate owners constitute a politically dominant class. Only when they see their power threatened by other classes of society (the industrial entrepreneurs, the urban middle sectors or even the peasants themselves through organized demands for land reforms) do they modernize their operations and use their resources more efficiently. They may then attempt to increase the exploitation of the peasantry or transform the semi-serfs into a rural proletariat, or simply evict them from their properties. In all of these cases social and political conflicts are likely to occur.

Estate agriculture represents a socially unjust and politically oppressive social system. Inequalities in wealth, income and social status between landowners and peasants are large and pervasive. Estate agriculture is always fraught with potential conflict, but it has also proved to be historically extremely stable, because it is tied to a fundamentally undemocratic and rigidly hierarchical social structure.

2.3.3 Modern Plantation Systems

These systems, also based on large landholdings as economic units, arose in the tropical areas for the production of commodities for export to the colonial metropolises or the industrialized countries. Plantations are commercial enterprises that rationalize their operations. Very often they are owned by foreign companies rather than individuals. They specialize in a single crop and frequently constitute veritable economic enclaves in the countries in which they operate. Their locally recruited labor force is not a traditional peasantry but a rural proletariat, working for a wage. Permanent plantation workers are often unionized and are able to engage in negotiations with management for higher wages, social security, fringe benefits and other issues. However, the seasonal workers come mainly from the peasant subsistence areas. Plantations are economic enterprises which require a high degree of organization, internal division of labor and specialization of tasks. They are more integrated into the international market than into the national economy in which they operate.

2.3.4 Family Farms

Family farms are the agricultural planner's utopian dream in the free-enterprise system. They are medium-sized, independent commercial enterprises, managed by an owner-operator at a relatively high level of technology and mechanization, with the occasional help of well-paid wage labor on a reduced scale, and provide the farm family with adequate income, giving it what might be termed 'middle- class status'. Family farms practice modem, rational agriculture and use their resources most efficiently. They sometimes combine different types of farming, rotate their crops, use fertilizers and improved inputs, and sell their produce on the market. Or else they specialize in cash crops with high unit value such as vegetables or flowers.

Family farms are not numerous in the underdeveloped countries for a number of reasons: the monopolization of the land in the hands of a few; the large number of traditional peasants who are unable to capitalize; the use of the land either for subsistence crops or for monoculture for export; the abundance of cheap under-employed labor; and the lack of integration between agriculture and industry within a strong internal market, which is one of the prerequisites for a family-farm economy in the industrialized countries.

Unless the traditional peasantry and the large mass of under- or unemployed agricultural laborers decrease sharply in the underdeveloped countries, it is unlikely that family farms will develop into a generalized kind of land-tenure system in the Third World.

The different kinds of agrarian structures mentioned above do not exist in isolation. Several of these systems may coexist within countries, depending upon a number of geographical, economic and historical factors. For example, in countries where European settlement took place at a relatively late date and where the native population was either exterminated or expelled from the settlement areas, family farms may have developed. In tropical areas where a native labor force was recruited during colonial times (or where slavery existed), plantation systems developed. In areas where a numerous peasantry was subordinated to a colonial system, the traditional large-estate system developed side by side with peasant holdings. Estates also existed in traditional feudal economies, such as those of the Middle East, where no foreign colonization took place. Peasant smallholdings, family farms, large estates, plantations and communal-tenure systems may exist within the same national society.

Often, the different systems are organically linked to each other, such as when plantations require labor from the areas of communal tenure (Africa) or when the large estate exchanges labor, produce and services with surrounding peasant holdings (Latin America).

The various systems use the resources at their disposal in different ways. It cannot be said that there exists a single optimal combination. Historical, political, social and institutional factors are as important as economic and technical ones. Small peasant holdings are usually considered inefficient in economic and technical terms. Their output per unit of labor is low. Their smallness makes the application of modern technology costly and impracticable. Yet in the absence of other employment opportunities, small peasant holdings use labor more intensively and their land and water resources more carefully. In contrast large traditional estates that monopolize the land in some countries are wasteful of their natural resources. Where they could modernize or mechanize, they prefer to use low-productivity labor. And when they do modernize, they often displace manpower, which, in a situation of large-scale unemployment, is socially and politically harmful. In the process of modernization of the large estates, the 'economic efficiency' of the production unit is frequently valued above the 'social efficiency' of the national economic system. We find still another combination in the communal or collective land-tenure systems associated with primitive shifting or slash-and-bum cultivation. In these systems, when the land-man ratio remains low, the tropical forest in which such cultivation takes place can regenerate itself over a period of several years. But when population pressure increases, or when deforestation takes place after a change in the use of the land, then the continued practice of shifting cultivation may rapidly destroy the remaining soil and thrust the primitive peasants into misery.

Land-tenure systems and agrarian structures are the result of historical development. While some may be the product of generations of spontaneous evolution, others were designed by governments or ruling elites with specific economic or political purposes in mind. They were not necessarily established for the maintenance of the ecological equilibrium; on the contrary, their evolution frequently leads to the breaking of the equilibrium, requiring new arrangements. Recent thinking about agricultural development has usually considered traditional peasant economies as existing prior, and being in a way opposed, to modern agriculture. Much has been written about how to transform traditional agriculture, how to modernize it. Different theories of economic growth foresee the gradual disappearance of peasant economies in the world. Some development theorists and planners believe that it is possible to transform traditional peasant plots into market-oriented, competitive family farms or enterprises, in imitation of what is supposed to have happened in the industrialized countries.[4] Other analysts see the process of capitalist development in agriculture producing on the one hand the concentration of wealth and resources in the hands of a new landlord or entrepreneurial class and on the other the progressive proletarianization of the dispossessed peasantry.[5]

[4] See: R Weitz, 1971: *From Peasant to Farmer, a Revolutionary Strategy for Development* (New York: The Twentieth Century Fund), for a forceful statement to this effect.

[5] See: R Stavenhagen, 1975: *Social Classes in Agrarian Societies* (New York: Anchor Books); and Keith Griffin, 1972: *The Green Revolution. An Economic Analysis* (Geneva: UNRISD).

2.4 Revival of the Peasant Economy

While a small number of entrepreneurial family farmers do indeed develop here and there out of the traditional peasant substratum of the underdeveloped countries, this is by no means a generalized tendency. A rural development strategy to this effect is doomed to failure in the sense that it may, to be sure, create a small middle class of family farmers in selected areas, but it cannot solve the problem of mass poverty in the rural areas. This can only be solved through an overall development strategy in which agricultural development is only a part.

The tendency towards economic polarization between a small landholding elite and a growing mass of proletarianized rural workers is clearly what is happening on a widespread scale in the underdeveloped countries. But contrary to predictions, even while this process is taking place, the traditional peasantry is not disappearing: on the contrary, it is in fact becoming more numerous in some areas.

The reasons for this are complex but it is essential to identify them for an understanding of rural poverty in the world today. We shall begin by defining peasant economy as the small-scale production of subsistence crops for local consumption by domestic groups based mainly on the use of family labor. For an economic characterization of peasant production, the legal aspect of land tenure is secondary: peasant production may take place on communally owned land, on private holdings, on leased or rented or sharecropped land, and on subsistence plots within large estates which peasants obtain in exchange for labor services.

Traditional peasants, as producers, are only loosely integrated into the capitalist system; their social world continues to be the local community with its own corporate structures, religious and political life, and cultural value systems. Peasants cultivate the land for their livelihood, rather than for monetary gain. Their lack of capital, of knowledge of the market, of formal education, and of opportunities is the result of their traditional subordination to local and regional power structures, in which the middlemen, the moneylender, the landlord, the political 'boss', all place insurmountable obstacles in the way of economic advancement and social improvement. Peasants are tied to their micro-plots, and unless large-scale institutional changes are brought about in the system which engulfs them, their transformation into independent, commercial, efficient farmers can be no more than wishful thinking.

Peasants are generally unable to capitalize. On the contrary, indebtedness is one of the more pervasive characteristics of peasant agriculture. Peasants cannot expand their operations, either because there is no more land available or because the price of land is too high (in both cases this may be so because of the monopolization of cultivable land by the regional landowner class), or because the amount of family labor available is limited and they lack the capital to employ wage workers.

Peasant farming, even while principally geared to the production of staple crops, is usually not able to satisfy the basic needs of the peasant household. With primitive technology and a small resource package, the peasant economy actually becomes increasingly decapitalized. If family labor were to be priced at prevailing wage rates (which it is not, in usual economic calculations, because it is an 'abundant' resource), the value of output is most likely to be inferior to the cost of the total inputs. In other words, the peasant farm is not only unable to turn a profit, it is often unable (in economic terms) to reproduce the labor force which is involved in its own production process.

The small peasant is placed before strong monopolistic elements in the rural land and capital markets. His industrial inputs, and of course his credit, are several times costlier for him than for the landlords or larger farmers.[6] Unable to keep his saleable surplus for long (owing to his constant need of cash for current consumption); he sells his produce at lower prices than the larger farmer. In other words, the peasant suffers a double squeeze. If to this is added the rent he pays, or the part of his crop he must deliver to the estate-owner or the sharecropping landlord, or the government tax, or interest on mortgage payments and so forth, we easily see how peasants are forced to transfer a part of their wealth to other sectors or classes of society. Thus their actual or potential surplus is skimmed off, or else they have to depress their already low living standards even further.

In these circumstances, why do peasants not simply give up their unprofitable activity and go into other sectors of economic life? Many of them do, and thus become proletarianized. But many of them do not, simply because the other sectors of the economy are unable to absorb them. Thus, in many areas of the world, peasants migrate temporarily to work in the modern agricultural sector, in the mines, in the cities, on construction sites and so forth. But they find neither stable employment nor adequate wages for themselves and their families. Time and again they are thrust back into subsistence agriculture only to be drawn again, temporarily, into wage work in the modern sector. The peasant economy has come to play the role of a labor reserve for capitalist enterprise in agriculture, mining and industry, as well as for the services sector.

In the underdeveloped countries, the modem agricultural, mining and urban-industrial sectors thrive by the use of cheap labor which the traditional peasant economies constantly provide. In the modern agricultural sector the need for labor is usually seasonal; but even in the other activities labor turnover is high and employment irregular. The modem sector is able to keep labor costs low not only by paying lower wages to migratory peasants than it would have to pay to a stable, permanent labor force, but also by not providing the various social services, housing, education and so forth which a permanent, stable labor force would be able to demand (particularly if it were unionized).

Economies exporting tropical commodities or raw materials are subject to severe international price fluctuations (sometimes artificially manipulated by the transnational corporations). When prices fall at short notice, cash-crop farmers and their laborers or sharecroppers, plantation workers, miners and other sundry workers directly or indirectly associated with the export economy are laid off. In

---

[6] See Griffin, *op. cit.*

the absence of viable employment alternatives, social security or unemployment compensation, they fall back upon the subsistence peasant economy for survival.

The peasant economy thus plays a dual role in the underdeveloped countries. On the one hand, however small and inefficient the peasant's plot, it serves to hold him on the land, thus lessening pressure on the non-agricultural economy in a situation of labor surplus. The peasant economy is able to reproduce the labor force at much lower cost to the economy as a whole than other sectors. It is thus in the interest of the modem, or capitalist, sector to maintain and, indeed, to re-create the peasant economy to a certain extent, as long as it remains subordinated to the needs of the modem sector. On the other hand, it provides a safety cushion for millions of underemployed workers who would otherwise openly starve (as many of them actually do in Africa and Asia), and who would generate enormous pressures on the social and political system.

Far from disappearing or receding into the background, the traditional peasant economy, linked to the modem capitalist economy through the various mechanisms that have been mentioned, turns out to be a major economic and social system in large parts of the world in the latter part of the twentieth century.

The world's peasantries are thus by no means marginalized or isolated vestiges of pre-capitalist economies. They cannot be written off simply because the theories of modernization or of capitalist development tell us that they should have disappeared long ago. It is among the peasantries in their various and complex manifestations that we find the largest numbers of those millions of rural poor which the World Bank has belatedly recognized as being a major challenge of our times.[7]

Strategies of development have generally by-passed the peasantry. They focus on the modern farmer, the agricultural entrepreneur, the so-called rural middle class. Even countries that have carried out land reforms do little, in the non-socialist world, for their peasantries once land has been redistributed. Rather, by simply distributing land and then concentrating additional efforts on those farmers 'most likely to respond' to monetary incentives, they are in fact re-creating the peasant economy. Mexico is a case in point: massive redistribution of land to the peasants during the 1930s; thereafter a thirty-year period of agricultural policies directed at strengthening the modem, entrepreneurial sector; the result being a considerable polarization of the agrarian structure with the concentration of wealth and resources among a small elite and the increasing marginalization of the large majority of subsistence peasants and landless workers.

[7] See: World Bank, 1975: *Rural Development* (Washington: World Bank).

2.5 The Peasant Household: Basic Economic Unit

An important fallacy appears to run through much of contemporary theorizing about rural development strategies. This is the emphasis placed on the farm as a self-sufficient enterprise. When the question of inputs, credit, market, technology, resources etc. is raised, this is usually done with respect to the farm unit as such, as if it existed within a social and institutional vacuum. The fact is, however, that in peasant economies the basic economic unit is not the farm at all, but the household. In peasant economies, as we have seen, farming is generally an uncertain and unstable occupation, and the peasant farm, whether it is devoted exclusively to subsistence crops or to cash crops, does not provide either sufficient employment or sufficient income to satisfy the basic needs of the peasant family (however these are defined).

The peasant household is not the characteristic nuclear family of urban settings, but frequently includes a fairly large number of members linked by kinship or affinity ties, covering various generations. Extended families, as these households or domestic groups are known in the specialized literature, are the real productive and consumption units of the peasant economy. Productive labor on the farm is but one aspect of a multitude of possible alternatives that the household actively pursues for its livelihood. The relative importance of direct farming depends, of course, on many local circumstances. The commitment may range from exclusive dedication (when no other alternatives are available) to a complementary activity (albeit a strategic one) when other alternatives present themselves.

The range of alternatives varies from country to country and from region to region, in accordance with the rate and kind of economic development that takes place at the national level. Thus, in many areas, temporary seasonal or pendular labor migrations are an essential complement to peasant farming. Elsewhere, or simultaneously, local handicraft production is a primary activity. This, however, is rapidly being displaced by the penetration of industrially manufactured goods even into the most remote areas, thus increasing the economic pressure on the peasant household. In still other areas, small-scale trade (sometimes even over long distances) is an essential source of much-needed cash. (Observers note the variety and colorfulness of market-places in western Africa or Indian Latin America, but seldom ask themselves about their economic function.) In many countries family members (male or female, usually the younger generation) seek employment in domestic or other services to supplement the peasant household's income; military service for the young men is another possibility. In some areas the development of international or national tourism opens up new vistas for local employment. (But it is generally underpaid and requires the supportive role of the peasant economy, cheap tourism being one of the attractions for the international jet-set who love to go to 'exotic' places.)

All of these activities cannot be accounted for simply as 'complementary income' for the peasant farm. They form an integral part of what we may call the peasant household's strategy for survival in underdeveloped capitalism. We must therefore attempt to understand the dynamics of the peasant household in its entirety as an economic and social unit. The role of family labor is paramount, as against the usual consideration of only the farmer or the head of the family as the visible economic pivot. Women, children and the elderly have important parts to play in the household's survival. As for children, their economic role frequently conflicts with their duty to attend school. When the men are away, women have to attend to the farm or the market-place. The internal division of labor in the household is essential to its economic function. Large families are of strategic importance. This is why birth-control programs so often run directly counter to prevailing cultural values among the rural population. These values are derived not only from some vague religious prejudice, but from the structural needs of the peasant economy.

Within this context, the time, energy and attention that the peasant household devotes to its plot of land are determined by two fundamental criteria: (a) the need for food; and (b) the available alternatives for obtaining monetary income. The relation between these two variables determines the nature and intensity of direct labor on the peasant farm. Contrary to facile references to the peasant's 'irrational behavior' or his abstract 'traditionalism', farm work is one of a number of carefully evaluated variables in the peasant household's economic calculations. At the level of subsistence living, a mistaken decision may make the difference between survival and starvation. The peasant household's margins for economic maneuver are slim, and the risks loom large.

Rural development strategies aimed at raising the standards of living of the rural poor must focus on the peasant household rather than on the peasant farm as such. This means that some of the basic premises upon which rural development planning has rested in recent decades should be rethought.

2.6 Objectives of Rural Development Strategies

Basically, different kinds of development strategies converge on a number of fundamental and common objectives. A clear understanding of these objectives is thus necessary for the adequate evaluation of different kinds of rural development strategies.

(1) Probably the most widespread objective at the present time is rapidly to increase agricultural output and productivity. The most spectacular advance in this field is from the various technical improvements known as the 'green revolution', that is, the various practices associated with the introduction of new, *high-yielding varieties* (HYVs) of seeds, mainly wheat, maize and rice. The 'green revolution' has had some success, especially in some Asian countries, in which it has contributed to considerable increases in agricultural output of basic grains in a relatively short time. Acreages covered with the new varieties of seeds have expanded rapidly. However, the 'green revolution'

has also run into some problems. The introduction of HYVs is associated with special technical and environmental factors (water for irrigation, fertilizers etc.), the success of which is in many areas reserved to a small, privileged class of richer farmers, who are also able to concentrate the benefits deriving from higher output. Generally, the small peasant has not adopted the new varieties. The 'green revolution', while contributing to the increase in output and productivity at the farm level, has also helped to aggravate income inequalities in the rural areas, and has increased the proletarianization of many small peasants.[8]

(2) Another overall objective of rural development strategies is to improve efficiency in the use of scarce land and water resources. Lack of consciousness about these matters has led to a dangerous depletion of soils in many countries. Millions of tons of good soil are washed away yearly by rains or floods or eroded by winds in mountainous or hilly areas. The haphazard cutting down of woods and forests has changed micro-climates and contributed to erosion. In other areas, the desert advances against the tropical rain forest or the cultivable areas. The control over soil erosion is closely linked not only to agricultural techniques, but also to the organization of production and the functioning of land-tenure systems.

The same may be said of the wastage and inefficient use of water. Many underdeveloped countries are partially arid and do not have favorable hydraulic resources. Certain kinds of irrigation systems, so necessary to increase agricultural production, are depleting ground-water deposits to levels at which their natural renovation is endangered. In other areas, water resources are contaminated through other uses with detriment to agriculture. This has even led in some cases to international conflicts. Water, like soil, is not inexhaustible, and agricultural planners have only recently become seriously concerned with these matters at international levels. The efficient use of water for irrigation is directly related to land tenure and the distribution and organization of farm units. It is thus a political and social as well as a technical problem.

(3) A serious obstacle to development in the Third World countries is the lack of capital resources. Agriculture is generally the last sector to receive new capital investments. In many countries, agriculture has actually been decapitalized. A more efficient use of capital resources is one of the principal objectives of many rural development strategies.

This is not an easy problem to solve nor are there any recipes to apply. Frequently, economic planners believe that any injection of capital will produce increased output, yet field studies and cost-benefit analyses of rural development projects in various parts of the world have shown that this is not necessarily so. On the contrary, massive investments in the rural areas have sometimes produced massive social and economic maladjustments. Too much

---

[8] See UNRISD, *op. cit.*

capital investment, and too rapidly, has led to tremendous wastages. Rural farm surveys have shown that whereas small peasant holdings are definitely undercapitalized, large modern estates or commercial farms may be highly overcapitalized. The modernization of agricultural operations has often led to the uncritical adoption of labor-saving mechanization, without making significant contributions to output. The efficient use of capital investments in the rural areas is not only a function of different factor availabilities, but also of the social organization of production, as well as the structure of the regional and national economy.

(4) Over the last decade it has become increasingly evident that one of the principal development objectives in the Third World countries must be the creation of employment opportunities for a growing mass of unskilled labor. Disguised unemployment is particularly acute in the rural areas, but detailed statistical information about this is difficult to come by. Satisfactory strategies for employment creation, particularly in rural areas, have not yet been devised. Many different measures are being considered: labor-intensive agricultural techniques, public works for infrastructure using manpower intensively, rural industrialization etc., combined with accelerated manpower training programs, the creation of regional poles of growth, the control of international transfers of technology, among others.

(5) Yet another objective is income redistribution. Economic growth over the last few decades has shown that aggregate and per capita output can be increased, but that the distribution of income between regions and social classes becomes more unequal. Agricultural growth has been no exception to this tendency. Modernization, mechanization, the 'green revolution' and other policies designed to further agricultural development have generally benefited a small group of large or richer farmers, merchants and middlemen. If a more equal or more just distribution of income (and with it, of social status and of political power) is indeed a development objective, then rural strategies must specifically design measures to implement this. The peasant farmer, the landless laborer, the migrant seasonal worker must be included in development plans, which is not always the case at the present time. Furthermore, in the underdeveloped countries, marketing and distribution networks tend to absorb a disproportionate part of rural and regional income. Agricultural development has furthered the growth of a 'rural bourgeoisie' whose increasing economic importance has only recently begun to be appreciated by students of the rural areas.

A rural development strategy aimed at improving the distribution of income would have to pay special attention to these questions, through the creation of marketing cooperatives or boards, state-owned purchasing and distribution agencies and other mechanisms allowing the rural producer easier access to urban and international markets.

(6) The final goal of a rural development strategy must, of course, be to raise the living standards of the rural population. Increased output and even increased monetary income do not automatically mean a better standard of living for the

peasantry in terms of material wellbeing, nutrition, education, security, leisure, mental health and social integration. All of these various goals require specific policies. Field studies in different parts of the world have shown that the sudden injection of money in a traditional economy may lead to wasteful spending, conspicuous consumption and produce socially harmful results. If increased output is to lead to real improvement in standards of living, in saving and productive investment, a number of social development policies must be carried out simultaneously with the introduction of economic measures on the production side. Education for consumption and better living is as important as training and incentives for increasing production. This requires the definition of collective rather than individual goals, of communal rather than personal improvements, of social rather than private interests. Specialists are not yet agreed as to what the relevant variables are, much less as to what are adequate indicators for measuring these variables. It is easier to measure increases in output than increases in social wellbeing.

The crisis of the world's agriculture and its peasant masses has led to the proposal of a number of development strategies in the rural areas, all of which have been tried with more or less success in different parts of the world.

2.7 Rural Development Strategies

2.7.1 Redistribution of Land

In areas of large estates and an oppressed peasantry, far-reaching agrarian reforms have redistributed the land to the peasants under various kinds of ownership arrangements. In some cases, the peasants have received small plots of land as proprietors; in others the land has been given to villages collectively, and heads of family have received individual usufruct rights to specific plots; in still others, cooperative or collective farms have been established on parts or on all of the old estate. In some cases the peasants have simply received title to the plot they have always worked, and only their labor services to the landlord have been abolished.

While the redistribution of land from the estate sector to the peasantry has everywhere had important political and social consequences (raising the social position of the peasant, making him a participant in political life), and has also allowed a rapid increase in the peasant family's consumption of foodstuffs (they can now retain more of their own produce, rather than transferring it to the landlord), the mere distribution of land rights or land titles does not solve the problems of agricultural backwardness and low incomes for the farmers. Land redistribution schemes must go together with a massive transfer of resources and inputs into the agricultural sector. Credit, technical assistance, supporting services of various kinds, must be channeled to the reform beneficiaries if substantive increases in agricultural output are to take place.

2.7.2 Abolition of Rents and Tenant Arrangements

Similar in effect to agrarian reforms in areas of large estates are measures designed to abolish rents and tenant arrangements for the benefit of the direct producer. Such policies do contribute to raise the level of income of the peasant, but they do not produce agricultural development by themselves unless accompanied by a whole series of additional measures. Their main result is a redistribution of agricultural income, at least for a period, before new kinds of exploitative structures (commercial or financial) again tie the peasant to some other social class that is able to extract surplus from his labor.

2.7.3 Landholding Reform

In regions where a traditional peasantry has been settled on the land for many generations, the landholding pattern be comes dispersed and complex. Commercial transactions, inheritance and other land transfers lead to the atomization of peasant property and to a crazy-quilt patchwork of tiny plots and parcels which is not conducive to the integration of viable economic units. Here, policies are put forward tending to consolidate dispersed peasant holdings, to redraw the local landholding maps and to create more stable and economically feasible farms. Again, unless these policies are accompanied by other measures, their beneficial effects may be short-lived.

2.7.4 Intensification of Peasant Agriculture

Where small peasant holdings are the result of land reforms, or where basic structural changes in the land-tenure system are not feasible, or in areas where a high level of unemployment characterizes the agricultural sector, thus requiring a part of the peasantry to remain on the land for several generations to come, policies leading to the intensification of peasant agriculture may be possible. This means channeling to the peasant the technical and financial assistance necessary to improve the use of his resources and the productivity of his labor, without necessarily changing the size of his farm. This means 'thinking small' rather than doing big things like building giant, expensive dams or introducing monster-sized tractors designed for wider open spaces and large private or collective farms. 'Thinking small' is not usually the way politicians or planners operate in the underdeveloped countries. Multipurpose dams, eight-lane highways and settlement schemes in faraway areas constitute more of a monument to statesmen preoccupied with their place in history, than do small irrigation networks built by local labor, soil-conservation projects of reforestation programs.

The intensification of peasant farming, designed mainly to increase the peasants' own income as well as to provide surplus produce for local and regional markets for the urban population, is not a 'popular' development strategy, because there has been so much emphasis on the backwardness and inefficiency of peasant agriculture that forward-looking planners want to do away with it altogether, and right away.

2.7.5 Family Farms

The development of family farms on the European or North American model has long been the purpose of many rural policymakers. The advantages of family farms are defended on economic and philosophical grounds. There is no doubt that, in certain social and economic environments, family farms are economically productive and competitive and able to absorb new technology productively, provide good incomes to their owners and contribute to the social and political stability of their countries (family farmers are usually conservative). But this rural development model is only possible in a situation where the labor force in agriculture has decreased to, say, less than 25 % of the total population, and where there is a dynamic internal market for agricultural products. Nowhere has it been possible to transform peasants into family-farmers, except on an experimental scale and at very high cost per unit (family or farm). Most land reforms in Latin America or Asia have not achieved the development of a stable class of numerous family farmers.

2.7.6 Cooperatives

Together with the development of the peasant economy or distributive land reforms or policies designed to further family farms, many strategies direct their attention to the growth and extension of various kinds of cooperatives of independent producers. Service, marketing, purchasing and credit cooperatives are well-established instruments that enable producers to reduce their costs and increase their incomes. The success of cooperatives depends on the economic solvency and stability of the members. But in underdeveloped countries in which there exist great income inequalities among the rural population, cooperatives generally benefit only the richer farmers and contribute to marginalize the lowly subsistence peasants, who might most benefit from cooperative arrangements.

2.7.7 Collective Farms

A final strategy of rural development consists in the furthering of different kinds of cooperative, collective or state farms. When a land reform takes place, large estates or plantations that are economically integrated units cannot be profitably subdivided into small plots or farms. Their maintenance as units may be necessary, even if the form of ownership or management changes. In such circumstances, there are strong arguments for state or collective management, on technical and economic grounds. In other cases, collective or state farms may result from the integration of small, individually owned units into larger ones.

The problems of state-owned or collective farms are many and well-known. They do not basically have to do with economic or technical rationality, but rather with psychological incentives, social organization and bureaucratic efficiency. In the underdeveloped countries, a strategy of collective farming seems to be increasingly envisaged by policymakers in order to confront the problems of increasing output, redistribution of income and creation of employment.

None of the aforementioned strategies needs to be taken by itself, even though policy-makers usually prefer to emphasize one or the other. It is possible that any one country may adopt one or several of these strategies of rural development simultaneously. The relative value of each strategy cannot be judged only on its own terms, but only in relation to the organization of production in agriculture at the local level. The viability of a rural development strategy depends on factors embedded in the wider socio-economic system. Each strategy has economic, legal, political and ideological implications which are beyond the scope of the rural planner or the agricultural specialist. A working knowledge of the political system is indispensable for a realistic appraisal of the possibilities of any one rural development strategy at any given time.

Recent experience has shown that there is no single rural development strategy applicable in all socio-economic and cultural environments. Unfortunately planners and policy-makers, for reasons of their own, often emphasize one strategy or one objective above all others (land distribution, or rural settlement, or the 'green revolution', or the creation of family farms etc.), and a country's scarce resources will go mainly into one channel. In rural development planning it is necessary to consider various objectives at the same time, and clearly to order them according to priorities. Frequently the priorities of urban-based national planners do not coincide with those of the rural population. Peasants are rarely consulted when development priorities are set. They should be.

Workshop on Southeast Asian indigenous peoples, Phnom Penh, Cambodia, 2008. *Source* Personal photographic collection of the author

With Inuit teaching staff at Arctic College, Nunavut, Canada, 2004. *Source* Personal photographic collection of the author

Chapter 3
Cultural Rights: A Social Science Perspective (1998)

Abstract The debates concerning the human rights of indigenous peoples and ethnic minorities have increasingly focused on cultural rights. UNESCO has made major contributions on these issues, related to cultural heritage, cultural diversity, intercultural education, indigenous knowledge, among other concerns. This chapter appeared in a publication by UNESCO in 1998.

3.1 The Problem of Cultural Rights

The *International Covenant on Economic, Social and Cultural Rights*, adopted by the General Assembly of the UN in 1966, makes only modest proposals regarding the latter. Article 15 mainly refers to the right of everyone to take part in cultural life, to enjoy the benefits of scientific progress and its applications, and to benefit from the protection of scientific, literary or artistic works. Article 13 posits the right of everyone to education, which "shall be directed to the full development of the human personality and the sense of its dignity". While cultural rights are also referred to in numerous international instruments as well as in several UNESCO conventions and recommendations, the full implications of cultural rights as human rights remain to be explored. This chapter aims to contribute to this debate from a social science perspective.

This text was first published as a chapter in: Halina Nieç (Ed.), 1998: *Cultural Rights and Wrongs* (Paris: UNESCO): 1–20. It is the revised version of an earlier paper: "Cultural Rights and Universal Human Rights", in Eide, Asbjørn et al. (Eds.), 1995: *Economic, Social and Cultural Rights* (Dordrecht, Martinus Nijhoff Publishers): 63–77. The permission to reprint this text was granted in July 2012 by UNESCO.

R. Stavenhagen, *Peasants, Culture and Indigenous Peoples*,
Texts and Protocols 4, DOI: 10.1007/978-3-642-34153-3_3, © The Author(s) 2013

For reasons which are self-evident, cultural rights are closely related to other individual rights and fundamental freedoms such as the freedom of expression, freedom of religion and belief, freedom of association, and the right to education. Cultural rights have not been given much importance in theoretical texts on human rights and, as Eide has pointed out, are treated rather as a residual category but states do have obligations to ensure the respect, protection and fulfillment of each one of these rights and these should be spelled out in the case of cultural rights and their various interpretations.

While some cultural rights can be dealt with exclusively within the framework of universal individual human rights, the relationship between culture and human rights is such that a broader approach is warranted. Portuguese that cultural rights, particularly those pertaining to the preservation of cultural heritage, the cultural identity of a specific people, and cultural development, are under circumstances considered as 'peoples' rights', and she calls for renewed efforts to frame such issues in international legal terms. In this essay I shall discuss some ideas concerning these issues.

If cultural rights are to be understood as any individual's right 'to' culture, then ideally there should not be any doubt as to the meaning of this term. Yet a cursory look at the way the concept 'culture' has been dealt with in some international documents and legal instruments shows a variety of usages. The right of a people to its own artistic, historical, and cultural wealth is stated in Article 14 of the Algiers Declaration on the Rights of Peoples, adopted by a non-governmental meeting of prominent experts in 1976. It has no legal standing in international law, not having been sanctioned by an inter-governmental body, but as Brownlie recognizes, it has had 'a certain influence', particularly to the extent that its ideas were reflected in the African Charter on Human and Peoples' Rights, adopted by the Organization of African Unity in 1981.

The right to the equal enjoyment of the common heritage of mankind is mentioned in Article 22 of the African Charter. The right to develop a culture has been asserted by UNESCO, and is mentioned in the African Charter as well as in the Algiers Declaration (Article 13). UNESCO also proclaimed a 'right to cultural identity' at the *World Conference on Cultural Policies* in 1982. Furthermore, the Algiers Declaration (Article 2) refers to the right to respect of cultural identity, and the right of a people not to have an alien culture imposed on it (Article 15).

The right of persons belonging to ethnic, religious or linguistic minorities to enjoy their own culture, to profess and practice their own religion, or to use their own language, in community with the other members of their group, is found in the *International Covenant on Civil and Political Rights* (ICCPR, Article 27.) This right was reaffirmed in the 1992 *UN De*claration on the Rights of Persons Belonging to National or Ethnic, Religious and Linguistic Minorities, which also calls upon States to take measures enabling persons belonging to minorities to develop their culture (Article 4). The Algiers Declaration refers to the right of minority peoples to respect for their identity, traditions, language, and cultural heritage (Article 19).

The *Genocide Convention*, adopted in 1948, defines genocide, which it declares to be a crime under international law, as the commission of certain acts "with intent to destroy, in whole or in part, a national, ethnical, racial or religious group, as such" (Article 2). Besides the actual killing of people, these acts include "causing serious bodily or mental harm to members of the group... forcibly transferring children of the group to another group ... etc.". Buergenthal rightly argues that by outlawing the destruction of national, ethnic, racial and religious groups, the Genocide Convention formally recognizes the right of these groups to exist as groups, which surely must be considered as the most fundamental of all cultural rights.

3.2 Underlying Conceptions of Culture

A careful reading of the above instruments will show that they refer indirectly to various distinct conceptions of culture which are not always clearly spelled out in the texts, and which are in fact often used rather loosely in general discourse. A systematic treatment of cultural rights as human rights will require a somewhat more rigorous conceptualization of cultural terminology.

3.2.1 Culture as Capital

One common view identifies culture with the accumulated material heritage of humankind in its entirety, or of particular human groups, including monuments and artifacts. According to this position, the right to culture would mean the equal right of access by individuals to this accumulated cultural capital. An extension of this view is the right to cultural development. Many governments as well as international organizations have established cultural development as a specific process of cultural change, which some people see as parallel or complementary to other forms of development, i.e. economic, political or social development.

The argument appears to be the following: if economic development means increasing goods and services, a rising GNP and better distribution thereof among the population, then cultural development would mean 'more culture' and better access to culture by more categories of people. Very often, this is interpreted as a purely quantitative process: the publications of more books, the establishment of libraries, wider circulation of newspapers and magazines, the building of museums, ownership of and access to television sets, and so on. The quantitative growth of cultural services is sometimes equated with the concept of cultural development, yet relatively little attention has been paid in official reports to the more qualitative dimensions of this process. What are the nature and the contents of such services? Can an increase in the number of TV channels really be equated with cultural development?

It is often assumed that there exists a consensus on what 'cultural development' is about. This is, however, a doubtful proposition. It may be argued, for instance, that many of the general statements about the 'right to cultural development'—implying more of the so-called cultural 'services'—too often hide the fact that there are underlying cultural conflicts in our societies, just as there are social, political, and economic ones. These conflicts occur over the recognition and identity of culturally defined groups in society, or about the nature of 'national' culture, or the aims of cultural policies. One widely accepted proposition is that there exists a 'universal' culture and that while some people are able to enjoy it, others may not have access to it. It follows that a right to culture should entail a more equitable access to this 'universal culture'.

This, however, is not the only possible approach, for the right to culture may also be interpreted as the right to a group's own culture, and not necessarily to some general or supposedly universal culture, because these two concepts are not necessarily coterminous. In fact, it has been pointed out repeatedly that so-called 'universal' culture is more often than not the world-wide imposition of 'Western' culture through the hegemonic practices of the Western powers, from the time of colonialism onwards. To be sure, UNESCO's efforts at universalizing the cultural heritage of humankind is a step away from the eurocentric tradition.

3.2.2 Culture as Creativity

A second widely held view does not regard culture necessarily as accumulated or existing 'cultural capital', but rather as the process of artistic and scientific creation. Accordingly, in every society there are certain individuals who 'create' culture (or, alternatively, who 'interpret' or 'perform' cultural works). Within this perspective, the right to culture means the right of individuals to freely create their cultural oeuvres with no restrictions, and the right of all persons to enjoy free access to these creations (museums, concerts, theatre, libraries etc.). Cultural policies are therefore directed to further the position of the individual cultural creator in society (the artist, the writer, the performer), and the right to the free cultural expression of these creators has become one of the most cherished human rights in contemporary times. The cultural creator, in fact symbolizes the freedoms of thought and expression, which has been one of the motivating forces of struggles for human rights throughout history. Let us simply remember the international outcry that occurs when artists or writers are banned, exiled or imprisoned (let alone executed) by authoritarian regimes.[1]

[1] Here Solzhenitzyn, Kurdish writers in Turkey, Salman Rushdie come to mind.

The view of culture as the result of the labor of cultural specialists has led to a widely held distinction between 'high' and 'low' culture. In Western countries, at least, cultural debates revolve around the relative weight and significance of 'elite' culture and 'popular' culture, the latter being defined as belonging to the sphere of the performing arts, usually channeled through the mass media and targeted at specific audiences by the cultural industries (for example: 'pop' music and 'pop' stars, 'cult' films, fashionable ways of dressing, youth culture promoted by highly paid and publicized promoters and performers). There is another view of popular culture which I will deal with below, but official policies directed towards the development of culture usually focus on 'elite' culture. In this case, cultural rights are easily identified with the rights of the cultural creators, the cultural specialists.

3.2.3 Culture as a Total Way of Life

A third view of culture comes to us from the discipline of anthropology. It takes culture to mean the sum total of the material and spiritual activities and products of a given social group which distinguishes it from other similar groups. Thus understood, culture is also seen as a coherent self-contained system of values, and symbols as well as a set of practices that a specific cultural group reproduces over time and which provides individuals with the required signposts and meanings for behavior and social relationships in everyday life.

The peoples of the world are the carriers of many thousands of distinct cultures. In some instances, all or most of a country's population share a common culture; in others, a state is made up of a variety of different cultures. There is no consensus about the actual number of existing cultures or about criteria for definition of membership (who belongs, who is excluded), though this is a crucial issue, particularly in relation to the problem of cultural rights. Similarly, there is no hard and fast way to draw a line distinguishing one culture from another. This is neither possible nor indeed necessary for our understanding of cultural dynamics. Generally speaking, specialists estimate that in contrast to the world's more or less 200 independent states, there are circa 10,000 distinct ethnic groups or ethnies, based mainly on linguistic differences, which is one of the main criteria, but by no means the only one, for distinguishing cultures from one another.

Cultures are not static. On the contrary, every identifiable culture is historically rooted and changes over time. Indeed, cultural change and the constant dynamic recreation of cultures is a universal phenomenon. A culture may be said to have particular vitality if it is capable of preserving its identity even as it incorporates change, just as a specific human being changes over time but retains her distinct identity.

There is, albeit, a danger in this approach, which is to treat culture as an object, a 'thing' which exists separately of the social space in which various social actors interrelate. Anthropology reminds us that the ethnic (cultural) identity of any group depends not so much on the content of its culture as on the social boundaries that define the spaces of social relationships by which membership is attributed in one or the other ethnic group.

Following from this critique, recent scholarship treats culture as something that is constantly constructed, reconstructed, invented and reinvented by ever-changing subjects; the emphasis here is on the way people perceive and speak about their culture, rather than on the culture itself (which by this criterion would have no objective existence outside of the individual's subjectivity). Customs and traditions are inherent elements of all observable cultures, yet traditions are constantly being invented and reinvented, and customs, by which people carry on their daily lives, regularly change to conform to varying historical circumstances, even as they strive to maintain social continuity. National cultures, which are so closely linked to state activity through governmental educational and cultural policies, are imagined collectively in historical process, and nations are sometimes described as 'imagined communities'. So while 'cultures' are given objective existence (people are born into a culture, social groups are identified by their cultures), they are also subjectively and variously constructed and fashioned by myriad individuals in continuing social interaction.

Why and how cultures persist, change, adapt or disappear, constitutes a special field of inquiry, and such questions are intimately related to economic, political and territorial processes. At any given time, in any given area, there may be majority and minority, dominant and dominated, hegemonic and subordinate cultural groups. *UNESCO's World Commission on Culture and Development* writes: "A country need not contain only one culture. Many countries, perhaps most, are multi-cultural, multi-national, multi-ethnic and contain a multiplicity of languages, religions and ways of living. A multi-cultural country can reap great benefits from its pluralism, but also runs the risk of cultural conflicts". The International Commission on Education for the Twenty-First Century argues that one of the problems of the future is "the multiplicity of languages, an expression of humanity's cultural diversity. There are an estimated 6,000 languages in the world, of which a dozen are spoken by over 100 million people".

While 'culture wars' (ideological tensions and conflicts over cultural issues such as education, language, cultural policies etc.) may occur in well-integrated societies without actually splitting them asunder (generally because other kinds of social, economic and political institutions help keep the contenders together), in other cases cultural issues have become powerful mobilizing forces in political strife around the world.

Consider just one instance among many, The Serbo-Croatian conflict which triggered the break-up of Yugoslavia, had much to do with long-standing rivalries between the national elites of the two republics over linguistic and religious issues. After decades of linguistic debates over the nature of 'Serbo-Croatian' or 'Croat-Serbian', in 1966 a large number of Croatian intellectuals published a Declaration insisting that Croatian was a distinct language and should be officially treated as such. One author writes that "the attempt to divide the languages was labeled nationalistic and was suppressed through a strong political campaign".

Other examples could be readily provided, but we should note that these are not exclusively cultural conflicts, but rather political ones over cultural issues. The way societies handle cultural differences among their populations may become highly politicized and these problems are often resolved at the political level.

3.3 Are Cultural Rights Culture Specific?

If culture is understood in this wider, anthropological sense, rather than simply as accumulated cultural capital or the product of the talents and labor of a small number of cultural creators, then it can be argued that cultural rights in their collective sense are culture-specific, that is, every cultural group has the right to maintain and develop its own specific culture, no matter how it is inserted or how it relates to other cultures in the wider context. This is now referred to as the right to cultural identity.

This approach raises a number of important issues regarding the right to culture. Basic to the Universal Declaration and the general instruments of human rights, is the principle of non-discrimination and equality. During the post-World War II debate on human rights, it was argued that if the principle of non-discrimination were strictly adhered to, then everybody would have equal access to all the 'goods' in the human rights basket, whether these are the civil and political rights or the economic, social, and cultural ones. Nonetheless, whether this is really sufficient to ensure the enjoyment of all of these rights by everybody remains a major question in the discussion of cultural rights.

It may be argued that the enunciation of the principle of non-discrimination is not sufficient within the framework and processes of present day societies to provide all individuals with equal access to all human rights. Moreover, even if true non-discrimination was a reality for everybody (which it is not), this would not necessarily ensure the enjoyment of specific cultural rights. A case can be made for the need to develop procedures and mechanisms for the affirmation and enjoyment of specific cultural rights of peoples; because unless such mechanisms are developed, cultural rights will not be fully enjoyed and guaranteed for everybody, notwithstanding the principles of equality and non-discrimination.

A second question which follows from the above is whether the concept of cultural rights can be adequately encompassed by a notion of universal individual rights, or whether they should be complemented by a different approach: that of collective or communitarian rights. There are persuasive reasons to include the latter approach. The principles of non-discrimination and equality, as set out in the *Universal Declaration and the Covenant on Civil and Political Rights* basically relate to the rights of individuals. However, when we refer to cultural rights, as well as to a number of social and economic rights, then a collective approach is often called for, since some of these rights can only be enjoyed by individuals in community with others and such a community must have the possibility to preserve, protect and develop its common culture. 'Cultural freedom', stated the Pérez de Cuellar report, "is a collective freedom. It refers to the right of a group of people to follow or adopt a way of life of their choice".

Beneficiaries of these rights may be individuals, but their content evaporates without the preservation and the collective rights of groups. Cultural rights pertain to persons belonging to specific cultures and shaped by these cultures, who engage in collective action, who share common values, and who can only be the bearers of these common values by joining with other members of their own group.

This line of reasoning necessarily poses the question about what kind of collectivities might be the logical subjects of such rights. Who are the bearers of these rights? Who are the actors, in sociological terms, that can claim these rights and to whom they are applicable? This is a complicated issue, because it leads directly into the discussion of the rights minority groups, cultural groups or peoples, concepts which do appear occasionally in international human rights instruments, but which are rarely adequately defined.

3.4 Cultural Diversity and Universal Human Rights

When we speak of cultural rights we need to take into account the cultural values that individuals and groups share, which they often hold dear and which shape and define their collective identities. The right to culture implies the respect for the cultural values of groups and individuals by others who may not share these values; it means the right to be different. How else are we to interpret the fundamental freedoms of thought, of expression, of opinion, of belief, that are enshrined in the *Universal Charter of Human Rights*?

Within this perspective on cultural rights, accommodation must be made for the fact that different cultures and civilizations do not necessarily share the same values. Perhaps many human values are held in common, but cultures may differ regarding others, as a result of different histories and social organization. While this holds true across state boundaries and civilizational fault lines, it also occurs within countries when culturally differentiated peoples share a common state and its territory.

But stressing the diversity of cultural values runs counter to the major thrust of human rights thinking in the world today, which holds the universality of human rights to be the basic underpinning of the international human rights edifice. Individual human rights must not only be universal in scope (that is, they apply to all human beings), but the underlying values must be universally shared. All human beings are equal; no matter what distinguishes them, they have the same rights. Yet, when we speak of the respect for different values as being essential to the concept of collective cultural rights, does that very distinctions not imply a rejection of universality in order to recognize the specificities of different social groups?

Whereas some scholars would deny the validity of this line of reasoning, arguing that cultural relativism jeopardizes the concept of human rights itself, there is no denying the fact that the real world is comprised of a multiplicity of culturally distinct groups and peoples. Unless the debate on cultural rights acknowledges the particular issues relevant to each cultural group, we may only be talking about meaningless abstractions.

This issue was recognized by the American Anthropological Association as early as 1947, when the *United Nations Commission on Human Rights* was still discussing various drafts of the Universal Declaration. At the time, the Executive Board of the AAA submitted a statement to the Commission, raising the question of how the proposed Declaration could be made to apply to all human beings. The Universal Declaration should not, said the American anthropologists, be conceived only in terms of the values prevalent in Western Europe and America. The Association argued, firstly, that the individual realizes his personality through his culture; hence respect of individual differences entails a respect for cultural differences.

Second, respect of differences between cultures is validated by the scientific fact that no technique of qualitatively evaluating cultures has been discovered.

Third, standards and values are relative to the culture from which they derive so that any attempt to formulate postulates that grow out of the beliefs or moral codes of one culture must to that extent, detract from the applicability of any Declaration of Human Rights to mankind as a whole.

Finally, the American Association of Anthropologists suggested that "only when a statement of the right of men to live in terms of their own traditions is incorporated into the proposed Declaration, then, can the next step of defining the rights and duties of human groups as regards each other be set upon the firm foundation of the present-day scientific knowledge of Man".

Thus, even as the Universal Declaration was being drafted a half century ago, American anthropologists considered it to be embodying the values of only one culture, and they questioned the automatic applicability of these standards to other cultures. In more recent years, particularly as African and Asian states joined the United Nations, this position has been taken up by many nations of the Third World, and it was highly visible at the World Human Rights Conference in Vienna in the summer of 1993.

The *African Charter on Human and Peoples' Rights*, for one, illustrates some of these difficulties. Article 17 takes from the Universal Declaration the stricture that "Every individual may freely take part in the cultural life of his community", and adds that "the promotion and protection of morals and traditional values recognized by the community shall be the duty of the State". Surely morals and traditional values are culturally defined, and to what 'community' does this article refer to?

Chapter II of the Charter refers not to rights but to duties. This is an interesting counterpoint to the question of rights. Article 29 states, inter alia, that the individual shall also have the duty "to preserve and strengthen positive African cultural values in his relations with other members of the society..." Now if to preserve and strengthen positive African cultural values is spelled out as a duty, then it may be assumed that there is a countervailing right to African cultural values. If it is every African's duty to strengthen and preserve these values, then every individual must have the right to enjoy them as well.

It should be noted that in its formulation, Article 29 distinguishes African values from non-African values. Secondly, it posits a certain unity or homogeneity of African values, since it makes no reference to possible internal diversity. And thirdly, if there are positive African values, then by implication there must be negative African or non-African values, which need not be strengthened or preserved. What, however, are these 'positive African values' and how are they defined? If this is a tough intellectual challenge, it is surely even more difficult to apply the concept legally. Just raising this problem means opening up a Pandora's Box of difficulties.

There is another dimension to the problem of cultural rights. We should be concerned not only about respect for variations of cultural values across international boundaries, between different regions, historical traditions and political systems, but within countries as well. Most of the states that signed the various international human rights instruments are themselves mosaics of different cultures. Whether these are the cultures of ethnic groups, minorities, nationalities or nations, in fact very few countries are culturally homogeneous. What does this diversity mean in terms of human rights and the right to cultural development? If we understand the right to cultural development to mean not only the right for individuals to innovate, to break new ground, and to receive more cultural services, but also the right to one's own culture–the culture of the group into which one is born, in which one lives and with which one identifies–, that is, the right to cultural identity, then the problem is, again, how are the objectives of cultural policies defined? When we speak of more and better education, what will the content of this education be? When we speak of cultural development, which cultures will be developed, and by whom?

We must perforce return to the question of cultural definitions, mentioned above. Over the past half century, development was frequently identified as a process of nation-building, an important aspect of which has been the development of a 'national culture', particularly in the countries of the so-called Third World, many of which achieved political independence during this time. Yet the conveniently ambiguous term of 'national culture' leaves open the question of whose nation and what kind of nation is to be developed. Connor has rightly suggested that the development of modern states has been more of a process of 'nation-destroying' than one of 'nation-building', in view of the fact that in the name of the modern nation-state numerous non-state peoples have in fact been destroyed or eliminated.

As the term has been used in recent history, nation-building generally implies a 'melting pot' of peoples, or else a process of 'national integration' or 'amalgamation'. This means that the various ethnic and cultural groups who for one historical reason or another find themselves living within the defined borders of an internationally recognized state, are expected to give up parts of their cultural identity to either adopt the values of the dominant or majority groups, or else to mix and create something entirely new (this is generally assumed to have been the process of nation-building in the United States). But usually it is the social groups who wield political power that determine the model to which national culture is to adhere, in other words, who decide the form and contents of educational and cultural policies.

Who are the people in power? On analysis, we may find that they often belong to one of the hegemonic cultural groups, who may be a majority or a dominant minority. And because they are the dominant group, they can define the national culture in terms of their own cultural identities. Hegemonic cultural groups who have the ability or power to define the national culture then expect all other groups to conform to this model, even if that means, in the long run, the destruction of other cultures. To cite but a few contemporary cases:

- The Sudanese state, controlled by the Arabic Islamic peoples of the north, attempts to impose Shari'a law and its own model of nationhood on the various peoples of southern Sudan, resulting in one of the longest civil wars in Africa. Peace talks began in 1997 but have not yet resolved the conflict. (Since this was written, Southern Sudan achieved its independence)
- The dominant Sinhala majority attempted to create Sri Lankan nationhood in its own image, provoking the emergence of the Tamil insurgency in that country in 1983, (Since this was written, the Tamil insurgency was defeated)
- The Turkish state has systematically denied cultural rights to the Kurdish minority, labeling the Kurds simply 'Mountain Turks', (Since this was written, Turkey has softened its position)
- Latin America's indigenous peoples were expected to conform to the 'national' culture developed by the mestizos and the ruling groups identified as the descendants of the Spanish colonial settlers. Recent indigenous social movements demand the right to cultural identity and territorial autonomy in some cases.
- The Fijian constitution of 1990 denied descendants of Indian immigrants the same citizenship rights as the native Fijians, but a revised constitution in 1997 redressed this imbalance.
- Malay nationhood is defined constitutionally by the politically dominant Malays, to the detriment of the Chinese community. While the Malays speak of a 'Malayan' Malaysia, the Chinese and other minorities would like to see a 'Malaysian' Malaysia, meaning the equal worth of all of its culturally distinct citizens.
- The unitary state of France does not formally recognize the existence of culturally distinct regional minorities on its soil (Bretons, Corsicans, Occitans…). The French Republic is identified as 'one and indivisible'.
- The French-speaking Québecois have been unable to convince other Canadians to recognize them as a 'distinct society' within the Canadian federation. (Since this was written Quebecois identity has been recognized more frequently)

The relationship between a cultural hegemon and other culturally distinct groups (whether referred to as peoples, nations or minorities of various types) is a complex issue that has serious implications for the definition and enjoyment of cultural rights. When a given ethnic group is able to extend its cultural hegemony over other, weaker groups, then it can safely be said that a violation of cultural rights occurs. In extreme cases, this has been labeled 'cultural genocide', but this notion is not actually referred to in the Genocide Convention or other legal human rights documents. More commonly, this process is referred to as ethnocide, and it occurs all over the world.

The attempts by hegemonic ethnic groups in control of the state to homogenize national culture, and resistance to such policies by the subordinate groups, is becoming the subject of international concern, and the issue has been taken up by the relevant United Nations bodies, such as the UN Sub-Commission on the Prevention of Discrimination and the Protection of Minorities. A growing number of states now recognize their multicultural heritage and some encourage the different groups on their territory to preserve and develop their separate cultures. Human rights discourse also refers to the right to be different. When we talk about cultural rights, we also mean the right of groups within a country to be able to maintain their own cultural identities, to be able to develop their own cultures, even (or especially) if these are distinct from the mainstream or dominant model of cultural development established by the so-called 'ethnocratic state'.

There have been bitter arguments between the 'universalists' and the 'contextualists' on these human rights issues, the former arguing that the Western liberal conception of human rights has universal validity, while the latter maintain that distinct cultures have different ways of dealing (or not) with human rights. In fact, the differences are not that insurmountable between the two extreme positions. Most political theorists now recognize that the most liberal and individualistic human rights policy, one that is neutral and impervious to any kind of cultural differences, in other words that is 'difference blind' so to speak, must nevertheless take such differences into account as being a sociological and often a political fact of life, when building a solid human rights edifice. To the extent that cultural identities are structured through collective interaction amongst socially and culturally defined individuals, it is clear that the respect for the individual rights of members of minorities or disadvantaged and marginalized groups must go hand in hand with the rights of such groups to preserve and develop their own identities. One political theorist, referring to Canada, argues that the liberal politics of individual rights must be expanded to include the politics of difference and of recognition.

Human rights policies are not entirely neutral because they are the result of the values shared by the majority or dominant culture in any given society at any one time. The fact that they are dominant does not necessarily make them universal. If culture is recognized as continuous practice—rather than a set 'thing'—then the evolution of human rights thinking in different cultural contexts should be seen as an ongoing process—a dialogic process, to be sure—rather than an 'either/or' scenario.

3.5 International Standards and Cultural Rights

The Universal Declaration does not mention minorities or any other human group, except the family. When the Universal Declaration was being drafted in the Human Rights Commission during the years 1946–1948, some states wanted to include specific provisions on cultural rights of minorities. However, the predominant view then was that this was not a general human rights issue, but relevant only to some specific, multicultural societies. Eleanor Roosevelt, the American chair of the Commission, explicitly stated this view: minority rights, she said, was a purely European matter which had no relevance to human rights in general. Due to the Commission's inability to achieve consensus on this point, the Universal Declaration deals with everyone's right to participate in and contribute to the cultural life of the community in a very general way, a statement which lends itself to various interpretations.

Understandably, several states and many individuals were dissatisfied by the way cultural rights and the rights of minority peoples were phrased in the Universal Declaration. But it is well to remember that at the same time that the UD was adopted, the General Assembly passed another lesser known resolution, in which it stated that "the United Nations cannot remain indifferent to the fate of minorities", and added that "it is difficult to adopt a uniform solution of this complex and delicate question, which has special aspects in each State in which it arises". It therefore asked the Sub-*Commission on the Prevention of Discrimination and Protection of Minorities of the Commission on Human Rights*, to devote some time to this question. After four decades of debates, reports and negotiations, a *Declaration on the Rights of Minorities* was finally adopted by the General Assembly of the United Nations in 1992.

One of the few concrete results of the earlier discussions in the United Nations on the question of minorities, is Article 27 of the *International Covenant on Civil and Political Rights*, which says: "In those states in which ethnic, religious or linguistic minorities exist, persons belonging to such minorities shall not be denied the right, in community with the other members of their group, to enjoy their own culture, to profess and practice their own religion, or to use their own language". This is the only article in the international bill of human rights that specifically addresses the question of the cultural rights of minorities.

This text may be considered as a step towards the recognition of the rights of cultural minorities, and perhaps even as a move beyond the abstract and universal consideration of individual human rights towards the idea of group rights. Nonetheless, some observers have commented that Article 27 falls far short of what is required in international instruments to ensure the protection of minorities and their cultural rights. Among others, the following shortcomings are underlined.

First, Article 27 begins with the statement: "in those states in which ethnic, religious, or linguistic minorities exist…" This leaves open the question of how to define what minorities exist in what states, and who defines them. Inasmuch as these international instruments are drafted and signed by states for their own use, the 'how' and 'who' obviously leaves governments free to determine whether their countries do or do not contain minorities. Often states, for their own political interests, deny that there are minorities within their borders, whereas minority groups wish to be recognized as such and demand their cultural rights. For example, Latin American states used to reject the idea that there were indigenous minorities in their countries, though this attitude has changed over the years. Turkey officially does not recognize the Kurds as a distinctive cultural group, calling them the 'mountain Turks'. The non-recognition of this cultural group has lead to severe human rights violations against those Kurds seeking recognition of their cultural identity. There are many other examples of states refusing to acknowledge the existence of minorities within their borders.

The second limitation is that Article 27 refers to persons belonging to minorities rather than to minority groups as such. The bearers of the right set out in the article are the individuals, not the groups. But it is obvious that such rights can only be enjoyed through the group to which the individual belongs. If the group is denied the right to its collective identity, then the individual's right is limited or denied.

A third problem is the passive wording of Article 27. It states that persons belonging to such minorities "shall not be denied the right…" Read literally, the Article does not establish any positive, affirmative right, or an obligation or duty on the part of states to carry out policies with the objective of developing these cultural rights. It simply enjoins the state from denying persons these rights.

Through interpretative practice, however, Article 27 has been given a more positive content, particularly in regard to rights of persons belonging to indigenous peoples. In several cases which it has dealt with, the Human Rights Committee of the United Nations, recognizes that Article 27 "includes certain economic and social rights of persons belonging to minorities, namely when such economic and social activities are essential to the culture of an ethnic community".

Notwithstanding this constructive interpretation, Article 27 is too weak a provision as a means of protecting and promoting the cultural rights of minorities since, without state interference, the general historical tendency is towards the destruction of minority cultures, through the structure of power relations in modern societies, the economic system, the impact of the mass media and the print medium, as well as common educational policies. Therefore, unless the rights of cultural minorities are taken seriously and mechanisms are developed by states and international organizations to actively promote, protect and strengthen minority cultures, they will be lost cultures even when there is no willful intention to destroy them. Unless positive steps are taken, we will indeed witness more and more 'nation-destroying' under the guise of 'nation-building' (that is, sociological 'nations' based on shared cultural identities will be superseded by political 'nations' co-terminus with the 'state'). This process, as mentioned before, has also been documented worldwide under the concept of 'ethnocide'.

The *UN Declaration on the Rights of Persons Belonging to National or Ethnic, Religious and Linguistic Minorities*, adopted by the General Assembly in 1992, takes a more positive view. Article 1 proclaims that "States shall protect the existence and the national or ethnic, cultural, religious and linguistic identity of minorities within their respective territories, and shall encourage conditions for the promotion of that identity".

However, the Declaration falls far short of ensuring the collective rights of cultural minorities. Article 4 speaks about measures to be taken by states "to create favorable conditions to enable persons belonging to minorities to express their characteristics and to develop their culture, language, religion, traditions and customs", in short, to express their identity. Yet it adds this restrictive caveat: "...except where specific practices are in violation of national law and contrary to international standards".

As we have seen, national law may at times be restrictive of the cultural rights of minorities, so this provision in Article 4 raises the issue of the relation between national law and international human rights standards, including the Declaration itself. Eide considers that

> the limitations set by national law, however, must not go beyond what is permissible under international human rights law. States cannot, by the use of national law, prohibit groups from developing their culture, unless the development is contrary to international standards. What is involved, in particular, is to prevent that 'development of culture' is used to maintain traditions which constitute violations of human rights, such as discrimination of women, imposed marriages, the maintenance of caste systems or other forms of systemic discrimination, female circumcision, or other forms of violations of international standards. It underlines the point, which is essential in all issues of accommodation that groups cannot demand to preserve those aspects of their culture and identity which are incompatible with universal norms.

The danger here is that some outside body might wish to set itself up as a judge of other peoples' cultures, a situation which recent historical process flatly rejects and which obviously contradicts the right of peoples to self-determination. Yet the issues raised by Eide are of the most crucial importance, inasmuch as they point to the inherent tension between universally accepted individual human rights and the collective rights of peoples and groups. A rule of thumb might be that from the standpoint of international human rights, individual human rights should have primacy whenever they are threatened by group rights (including cultural rights). It is clear, however, that there is no universal consensus about this problem. Habermas argues cogently that "a correctly understood theory of rights requires a politics of recognition that protects the integrity of the individual in the life contexts in which his or her identity is formed".

A case in point that has received widespread attention in recent years relates to the rights of indigenous peoples, defined as the descendants of those populations that inhabited a given territory before the arrival of a conquest or settler society, to which they were thereafter subordinated. In contrast to nations who achieve or regain their political sovereignty in later years (decolonization), indigenous peoples the world over, who have often suffered severe discrimination and marginalization, and frequently been denied full citizenship, demand not only equal rights with all other citizens, but also the recognition of their own collective identities, including cultural identity, social organization, territorial links, and inclusion as equal partners in the wider society. Collective cultural rights are an important part of indigenous group claims. Some states have made progress in recognizing these rights (Bolivia, Canada, Australia, Norway); others, however, (ex., Brazil, Mexico, India) resist recognizing such rights and insist rather on the concept of the all-inclusive 'civic' nation that rejects the legal and political recognition of sub-national collective identities.

Convention 169 of the *International Labour Organization* (ILO) is one of the few international legal instruments that refers specifically to indigenous peoples. A draft Declaration of Indigenous Rights (including cultural rights) is being discussed in the UN Human Rights Commission and is expected to be adopted by the General Assembly before the end of the World Decade of Indigenous Peoples in 2004. The *Organization of American States* (OAS) is considering a similar declaration for the Americas. While indigenous organizations wish to be called 'peoples', many states reject this terminology because of its implications in international law (namely, the right of peoples to self-determination, that states are not willing to concede to indigenous populations, or to minorities for that matter). Indigenous peoples' rights are essentially cultural rights.

3.6 Cultural Rights and State Policies

How do states deal with these issues? In most countries where minorities exist, state policies are designed to assimilate or integrate minorities into the prevailing model of the national culture. In some cases, this might be a shared objective. For example, in immigration states, where people come from various parts of the world, the immigrants may actually want to shed their traditions and become part of the new 'melting pot'. However, even in societies that sustained the idea of the 'melting pot' for many generations, this ideal has increasingly come under criticism. Too often, policies of national integration, of national cultural development, actually imply a policy of ethnocide, that is, the willful destruction of cultural groups.

Ethnocide is distinct from genocide, which is the physical destruction of peoples, but it is equally reprehensible. When the Genocide Convention was discussed in the United Nations, there was considerable debate about the need to define 'cultural genocide', but the matter was not pursued because of the difficulties involved. Today, the concept of ethnocide has come to be accepted as the definition of a process of deliberate cultural destruction, although the concept has not yet been incorporated into any international legal instrument.

Now, if there is ethnocide, then one could say that there might be a right to 'counter-ethnocide' through 'ethnodevelopment', that is, policies designed to protect, promote and further the culture of distinct non-dominant ethnic groups within the wider society, within the framework of the nation-state or the multinational state. Ethnodevelopment might be an aspect of the 'right to development' which the United Nations General Assembly proclaimed in 1986.

The cultural development of peoples, whether minorities or majorities, must be considered within the framework of the right of peoples to self-determination, which by accepted international standards is the fundamental human right, in the absence of which all other human rights cannot really be enjoyed. Let us recall that Article 1 of both the international human rights covenants establishes in identical terms the right of peoples to self-determination. The international community is at odds as to who actually possesses the right to self-determination, and what the right entails in different contexts.

It is generally assumed that the populations of non-self-governing territories hold the right to decide whether they wish to become independent states or not, just as the latter have the right to maintain their independence. This is known as 'external self-determination'. For ethnic and cultural groups inside sovereign states, however, the issue of self-determination is a different one; but for exceptional circumstances, international law does not recognize the right of self-determination to minorities within independent states, if this is understood as secession or establishing an independent state of their own.

Nevertheless, there is increasing support for the view that minorities have a right to internal self-determination, which is less territorial than cultural; to maintain and preserve their separate identities within the larger national society, sometimes within a framework of autonomy. It remains a subject of intense controversy, however. This is mainly because governments fear that if minority peoples hold the right to self-determination in the sense of a right to full political independence, then existing states might break up through secession, irredentism or the political independence of such groups. State interests thus are still more powerful at the present time than the human rights of peoples. Some states, indeed, use the argument of 'cultural relativism' to weaken, when not actually to repress, human rights within their jurisdiction.

3.7 Indigenous Peoples: A Case for Cultural Rights

The struggle for the rights of indigenous peoples illustrates well some of the issues discussed here. I shall refer briefly to the 40 odd million indigenous of the Americas, particularly in the Latin American area, where over 400 distinct groups have been identified. The United Nations defines indigenous peoples as follows:

> Indigenous communities, peoples and nations are those which, having a historical continuity with preinvasion and pre-colonial societies that developed on their territories, consider themselves distinct from other sectors of the societies now prevailing in those territories, or parts of them. They form at present non-dominant sectors of society and are determined to preserve, develop and transmit to future generations their ancestral territories, and their ethnic identity, as the basis of their continued existence as peoples, in accordance with their own cultural patterns, social institutions and legal systems.

Since 1982, the Working Group on Indigenous Populations of the *UN Sub-Commission on Prevention of Discrimination and Protection of Minorities*, has been drafting a *Declaration of the Rights of Indigneous Peoples* which is to be submitted to the General Assembly before the end of the International Decade of Indigenous Peoples. In the drafting process numerous indigenous organizations from all over the world took part together with government representatives. Cultural rights figure prominently in this document, such as the right of indigenous peoples to be protected against cultural genocide and ethnocide; and "the right to the restitution of cultural, intellectual, religious and spiritual property taken without their free and informed consent or in violation of their laws, traditions and customs". The intellectual property rights of indigenous peoples include "sciences, technologies and cultural manifestations, including human and genetic resources, seeds, medicines, knowledge of the properties of fauna and flora, oral traditions, literature, designs and visual and performing arts", which the UN Sécial Rapporteur describes as their 'cultural heritage'. It will be appreciated that all of the above fall within the scope of what we have called collective cultural rights.

Whereas the UN draft declaration on the rights of indigenous peoples does not yet constitute an international legal document, this is not the case of Convention 169 of the International Labour Organization, which has been ratified by a number of states since its adoption by the ILO General Conference in 1989. The Convention stipulates that indigenous "peoples shall have the right to retain their own customs and institutions, where these are not incompatible with fundamental rights defined by the national legal system and with internationally recognized human rights".

Struggles for indigenous rights occur under the umbrella of these international documents, which provide a framework for more specific regional human rights instruments and for new legislation that has been undertaken at the national level in a number of Latin American countries. For centuries, indigenous peoples were oppressed, exploited and marginalized under a strict colonial regime in Latin America. During the independent period (since the early nineteenth century), the indigenous underclass (consisting mainly of poor subsistence peasants, migrant workers and serf-like rural laborers), were neglected and ignored by the ruling classes in their vision of building a modern nation-state. These states were built upon the backs of Indian labor, and even though Indians enjoyed formal equality as citizens in some countries, they lacked most of the attributes of full citizenship. During the present century government policies attempted (with some success) to assimilate and incorporate Indian populations into the (non-Indian) mainstream. Still, Indian identities survived, persisted and, in some cases, even thrived. In Bolivia and Guatemala, Indian populations constitute a demographic majority. In other countries, such as Ecuador, Mexico and Peru, they make up significant minorities, especially in certain regions in which their demographic density is high. Indians are no longer only a rural population: economic changes and massive migrations have brought them increasingly into the large metropolitan areas, where indigenous identities are undergoing rapid transformation.

Under these circumstances it was only a matter of time before indigenous peoples began to organize themselves socially and politically in order to demand their basic rights, challenge established government policies, and claim adequate representation in the political process. The organizing process began in the 1960 and 1970s, and within two decades indigenous peoples have emerged as new social and political actors in Latin America. Their claim to cultural rights has become an important mobilizing principle: recognition of their identities, public use of their languages, bi-lingual and multi-cultural education, access to mass media, protection of their intellectual property and cultural heritage, control over their natural resources, respect for their traditional social and political organization, recognition of their customary legal systems within the wider framework of national law.

Among the more insistent claims of indigenous organizations in Latin America are the right of peoples to self-determination and the right to autonomy. Both rights are interrelated. The *Vienna Declaration on Human Rights* (1993) speaks of indigenous 'people' rather than 'peoples', and this semantic difference is of considerable political importance to indigenous organizations. International law does not accord the right to self-determination to indigenous peoples (nor does it to minorities), and therefore states discourage the use of the term 'indigenous peoples' in international legal instruments. For the opposite reasons, indigenous organizations tend to insist upon it. All these are issues which are being fought over politically in Latin America at the present time, and in these controversies Indian intellectual and political actors play an increasingly important role.

Argentina, Bolivia, Brazil, Colombia, Nicaragua and Paraguay are among the countries that have adopted new constitutions or constitutional amendments since the early 1980s which include references to indigenous cultural rights. Other states have enacted modest or far-reaching legislations concerning their indigenous populations. In 1996, a strenuously negotiated peace agreement put an end to a 30 year old civil war in Guatemala in which the majority indigenous populations were either victims or active participants. The cultural rights of indigenous peoples are a crucial element of this peace agreement. In Mexico the indigenous peasants of Chiapas who rose up in arms against the national government in 1994 advanced claims for autonomy and cultural rights in the peace negotiations which had not yet concluded by the middle of 1998. There are numerous local and regional experiences in which one or several collective cultural rights are being implemented in some Latin American country or other, particularly in the field of language and educational policies. Things are more complex and difficult regarding territorial rights, control over natural resources, or intellectual property and customary law.

3.8 Towards Multicultural Citizenship

As Latin American nations struggle to redefine their relations with indigenous peoples and thus to redefine themselves, the idea of cultural or multicultural citizenship has become a useful concept. By this I mean the recognition of indigenous people qua peoples with their own legal status and the right to self-determination; indigenous communities as subjects of public law with autonomic rights, indigenous languages as national languages, the demarcation of their own protected territories, the right to the management of their resources and their development projects, respect for their internal norms of local government and their customary legal systems, cultural and religious freedom within the community, as well as political participation and representation at the regional and national levels. It is not only al question of ensuring individual and collective rights within the existing state structures, but of redefining the very notion of state and nation. Cultural citizenship as far as indigenous peoples are concerned, should have two essential points of reference: the unity of the democratic state and the respect for individual human rights within the autonomic collectivities and units that may be established. Neither pure individualistic liberalism nor the corporativist structure of a centralist state (such as exists in Latin America) satisfy the requirements of multicultural citizenship: this can only be achieved through democratic practice, dialogue, tolerance and mutual respect.

It is within the framework of some of the issues set out above that the argument for the development of cultural or multicultural citizenship has emerged as a constructive approach to the cultural rights of groups within the modern nation-state. The anthropologist Renato Rosaldo and his colleagues, for example, argue for the need for cultural citizenship among the Hispanic populations in the United States, meaning thereby the reclaiming and reconstruction of the social and geographic spaces of Latino communities. "Cultural citizenship", he writes, "operates in an uneven field of structural inequalities where the dominant claims of universal citizenship assume a propertied white male subject and usually blind themselves to their exclusions and marginalization's of people who differ in gender, race, sexuality, and age. Cultural citizenship attends not only to dominant exclusions and marginalization's, but also to subordinate aspirations for and definitions of enfranchisement". Furthermore, cultural citizenship means empowerment, "a process of constructing, establishing and asserting human, social and cultural rights". It is to be thought of as "a broad range of activities of everyday life through which Latinos and other groups claim space in society and eventually claim rights.... [it] allows for the potential of opposition, of restructuring and reordering society". In a similar vein, the Peruvian historian Rodrigo Montoya suggests that 'ethnic citizenship' be recognized to indigenous peoples who wish to use their own language and reproduce their own culture within the wider society.

From the perspective of political theory, Kymlicka argues for a form of differential citizenship in multicultural societies such as Canada, where different identities should be accorded recognition not in order to fragment a fragile nation, but on the contrary in order to integrate and strengthen it. "Shared values [he suggests] are not sufficient for social unity...The missing ingredient seems to be the idea of a shared identity... [which] derives from commonality of history, language and maybe religion. But these are precisely the things which are not shared in a multination state". Kymlicka calls upon liberal states to "insure that there is equality between groups and freedom and equality within groups". Similarly, the philosopher Charles Taylor holds that "liberalism can't and shouldn't claim complete cultural neutrality". He proposes a politics of 'mutual recognition' in multicultural societies. Also within the liberal perspective, Spinner pleads for a new form of 'pluralist integration' of culturally distinct groups in the United States.

The cultural rights of distinct ethnic groups in existing states can thus be considered under the framework of ethnonationalist struggles or within the approach of liberal citizenship. Whether group claims for cultural rights fall into one or the other of these camps depends upon particular circumstances, and they are not necessarily mutually exclusive. Pluralism is but one of a number of policies that might satisfy claims to cultural distinctiveness. Multicultural citizenship constitutes another frame that promises greater empowerment and participation by disadvantaged collectivities, even as nation-states are challenged to reconsider their traditional and often legally enshrined perceptions of themselves.

Rolfo Stavenhagen with indigenous women at a meeting in Salekard, Nemets, Russia, 2008. *Source* Personal photographic collection of the author

Masai boys in Tanzania, wondering about what the future may hold for them, 2007. *Source* Personal photographic collection of the author

Indigenous children at school in Thailand. *Source* Personal photographic collection of the author

Bibliography

American Anthropological Association, 1947: "Statement on Human Rights", in: *American Anthropologist* 49, 4 (October-December): 539–43.

Anderson, Benedict, 1983: *Imagined Communities. Reflections on the Origin and Spread of Nationalism* (London: Verso).

Barth, Frederick, 1969: *Ethnic Groups and Boundaries* (Oslo: Universitetsforlaget).

Bloch, Anne-Christine, 1995: "Minorities and Indigenous Peoples", in: Eide, A., C. Krause and A. Rosas (Eds.): *Economic, Social and Cultural Rights* (Dordrecht: Martinus Nijhoff Publishers).

Brölmann, Catherine et al. (Eds.), 1993: *Peoples and Minorities in International Law* (Dordrecht: Martinus Nijhoff Publishers).

Brownlie, Ian, 1988: "The Rights of Peoples in Modern International Law", in: James Crawford (Ed.): *The Rights of Peoples* (Oxford: Clarendon Press).

Buergenthal, Thomas, 1988: *International Human Rights* (St. Paul, Minnesota: West Publishing Co).

Burgers, J. Herman, 1990a: "The Function of Human Rights as Individual and Collective Rights", in: J. H. Burgers et al. (Eds.): *Human Rights in a Pluralist World. Individuals and Collectivities* (London: Meckler): 63–74.

Burgers, J. Herman, 1990b: "The Right to Cultural Identity", in: J. H. Burgers et al. (Eds.): *Human Rights in a Pluralist World. Individuals and Collectivities* (London: Meckler): 251–253.

Connor, Walker, 1972: "Nation Building or Nation Destroying", in: *World Politics*, 24, 3.

Donnelly, Jack, 1989: *Universal Human Rights in Theory and Practice* (Ithaca: Cornell University Press).

Eide, Asbjørn, 1994: *Peaceful and Constructive Resolution of Situations Involving Minorities* (Oslo: Norwegian Institute of Human Rights).

Eide, Asbjørn et al. (Eds.), 1995: *Economic, Social and Cultural Rights* (Dordrecht, Martinus Nijhoff Publishers).

Ermacora, Felix, 1964: *Der Minderheitenschutz in der Arbeit der Vereinten Nationen* (Wien-Stuttgart: Wilhelm Braumüller).

Ermacora, Felix, 1984: "The Protection of Minorities before the United Nations", in: Academy of International Law (Ed.): *Recueil des Cours*, vol. 182 (The Hague: Martinus Nijhoff Publishers).

Flores, William V. and Rina Benmayor, 1997: "Constructing Cultural Citizenship", in: *Latino Cultural Citizenship. Claiming Identity, Space and Rights* (Boston: Beacon Press).

Habermas, Jürgen, 1994: "Struggles for Recognition in the Democratic Constitutional State", in: Charles Taylor et al. (Eds.): *Multiculturalism* (Princeton: Princeton University Press).

Hannum, Hurst, 1992: *Autonomy, Sovereignty, and Self-Determination. The Accommodation of Conflicting Rights* (Philadelphia: University of Pennsylvania Press).

Hobsbawm, Eric, and Terence Ranger (Eds.), 1983: *The Invention of Tradition* (Cambridge: Cambridge University Press).

Huntington, Samuel P., 1997: *The Clash of Civilizations and the Remaking of World Order* (New York: Simon & Schuster).

Kymlicka, Will, 1995: *Multicultural Citizenship* (Oxford: Clarendon Press).

Montoya, Rodrigo, 1996: "La ciudadanía étnica como un nuevo fragmento en la utopía de la libertad", in: González Casanova, Pablo and Marcos Roitman Rosenmann (Eds.): *Democracia y estado multiétnico en América Latina* (México: UNAM).

Neçak Luk, Albina, 1995: "The linguistic aspect of ethnic conflict in Yugoslavia", in: Payam Akhavan and Robert Howse (Eds.): *Yugoslavia, the Former and Future* (Geneva: UNRISD).

Pollis, Adamantia, 1996: "Cultural Relativism Revisited: Through a State Prism", in: *Human Rights Quarterly*, 18,2.

Preis, Ann-Belinda S., 1996: "Human Rights as Cultural Practice: An Anthropological Critique", in: *Human Rights Quarterly*, 18,2.

Prott, Lyndel V., 1988: "Cultural Rights as Peoples' Rights in International Law", in: J. Crawford (Ed.): *The Rights of Peoples* (Oxford: Clarendon Press): 92–106.

Rosaldo, Renato, 1997: "Cultural Citizenship, Inequality and Multiculturalism", in: Flores and Benmayor (Eds.): *Latino Cultural Citizenship* (Boston: Beacon Press).

Spinner, Jeff, 1994: *The Boundaries of Citizenship. Race, Ethnicity, and Nationality in the Liberal State* (Baltimore: The Johns Hopkins University Press).

Stavenhagen, Rodolfo, 1990a: *The Ethnic Question. Conflicts, Development and Human Rights* (Tokyo: United Nations University Press).

Stavenhagen, Rodolfo, 1990b: "The Right to Cultural Identity", in: J. H. Burgers et al. (Eds.): *Human Rights in a Pluralist World. Individuals and Collectivities* (London: Meckler): 255–258.

Stavenhagen, Rodolfo, 1993: "Self Determination, Right or Demon?", in: *Stanford Journal of International Affairs*, II,2: 1-12.

Stavenhagen, Rodolfo, 1995: "Indigenous Peoples: Emerging Actors in Latin America", in: *Ethnic Conflict and Governance in Comparative Perspective*. Working Paper No. 215 (Washington, D.C.: Woodrow Wilson International Center for Scholars).

Stavenhagen, Rodolfo, 1996: "Indigenous Rights: Some Conceptual Problems", in: Elizabeth Jelin and Eric Hershberg (Eds.): *Constructing Democracy: Human Rights, Citizenship and Society in Latin America* (Boulder, Westview Press).

Taylor, Charles et al., 1994: *Multiculturalism* (Princeton: Princeton University Press).

Thompson, Michael, Richard Ellis and Aaron Wildavsky, 1990: *Cultural Theory* (Boulder: Westview Press).

Tomuschat, Christian (Ed.), 1993: *Modern Law of Self-Determination* (Dordrecht: Martinus Nijhoff Publishers).

UNESCO, 1982: *World Conference on Cultural Policies. Final Report* (Paris: UNESCO).

UNESCO, 1995: *Our Creative Diversity. Report of the World Commission on Culture and Development* (Paris: UNESCO).

UNESCO, 1996: *Learning: The Treasure Within. Report to UNESCO of the International Commission on Education for the Twenty-first Century* (Paris: UNESCO).

United Nations, Department of Public Information, 1984: *The United Nations and Human Rights* (New York: United Nations).

United Nations, Department of Public Information, 1987: *Is Universality in Jeopardy?* (New York: United Nations).

Chapter 4
Exclusion and Human Rights (2000)

During the nineties, a number of specialized agencies of the United Nations became increasingly concerned with human rights issues and the widespread poverty around the world, a concern that led to a number of international forums and research projects devoted to these topics. At one such meeting, organized by the *United Nations Research Institute for Social Development* (UNRISD) I presented this paper on the interagency between poverty, marginality and human rights, with special emphasis on indigenous peoples. (This paper was presented at the United Nations Research Institute on Social Development Conference on "Racism and Public Policy", Durban, South Africa, 3–5 September, 2001. This paper is in the public domain.).

Abstract Noting that globalization has produced both winners and losers, this chapter argues that poverty and inequality lead to social exclusion and discrimination of large numbers of people in the world, making them vulnerable to massive abuses of human rights. This has been the case particularly among rural folk, linked to unjust agrarian structures, most notably in the Third World countries. As a glaring example of the structure of injustice, the situation of Latin America's indigenous peoples is examined. Throughout their history since the colonization of America, the indigenous have been exploited, oppressed and discriminated against. In modern times, governmental policies attempted to assimilate them into the larger society, thereby violating their cultural and collective rights. In recent decades, emerging indigenous organizations have struggled for recognition of their rights as peoples, which have been progressively accepted in national legislations and in international human rights instruments.

4.1 Globalization and Injustice

There is much talk nowadays about how a single global market will provide the framework for growth and development worldwide well into the next century, yet it is clear that there are serious disparities and vast gaps between the haves and the have-nots in the world economy. While much attention is being paid to the transnational factors of growth, relatively less concern is expressed about the losers in this new planetary game. Moreover, it is not too farfetched to argue that the economic losers become the socially excluded, and that it is here, among this vast category of the world's population that we find the major obstacles to the effective implementation of human rights. Poverty–marginality–human rights abuses are the three principal terms of a social equation that is emblematic of the beginning of the new millennium. The massive protest staged by non-governmental organizations of all

R. Stavenhagen, *Peasants, Culture and Indigenous Peoples*,
Texts and Protocols 4, DOI: 10.1007/978-3-642-34153-3_4, © The Author(s) 2013

kinds at the World Trade Organization conference in Seattle in December 1999 was the expression of growing worldwide dissatisfaction with the globalized hegemonic neoliberal economic model.

In its 1995 *World Economic and Social Report*, the United Nations Organization reports that world economic growth appears to have reached a sustainable 'cruising speed' of 3 %/year, but it recognizes that the product is not equally distributed among regions and nations. When the analysis focuses not on countries with their aggregate indicators (such as GNP or per capita income), but rather on human beings, then a wholly different and worrisome picture emerges. The United Nations Human Development Report, for instance, states that during the past three decades global income disparity has doubled: the richest 20 of the world's people now receive more than 150 times the income of the poorest 20 %. The UN also concludes that there is no automatic link between income and human development, and when income distribution is factored in, the Human Development Index (HDI) of a number of countries falls sharply (United Nations 1993).

Worldwide concern over increasing poverty led to the World Social Development Summit in Copenhagen in the spring of 1995, where attending chiefs of state restated their commitment to combat poverty, work towards the creation of fuller employment and promote social integration. A study prepared for the Summit by the United Nations Research Institute on Social Development points to the painful social consequences of the widely fashionable structural adjustment policies imposed on numerous countries by the international financial agencies such as the World Bank and the International Monetary Fund. According to UNRISD (1995), these are not only short term 'adjustment costs' but rather long term tendencies which actually endanger the potential benefits of the economic adjustment policies themselves. During the expansive eighties, for example, real minimum wages dropped by 20 % in most African countries and by approximately 50 % in Latin America. In this region, the total number of poor (whose income is below sixty US dollars a month) increased by 60–196 million between 1980 and 1990. Those Latin Americans who live in extreme poverty (with incomes of <30$ a month) grew from 19 to 22 % of the total population (94 million people), which means that one in five Latin Americans does not dispose of sufficient income to satisfy her basic human need in food intake (CLCDS 1995).

The vast majority of the world's poor are found in the so-called developing economies, and specifically in the rural areas. Indeed, agricultural activities are still the largest single occupation in the world (more people attempt to make a living from agriculture than any other single source). Except in the industrialized countries, agriculture everywhere employs from one fifth to over one half of the labor force (Asia, Africa, Latin America). Still, in most areas the share of agricultural production has decreased within the gross domestic product and except for a relatively small number of success stories, overall agricultural productivity has fallen and is expected to decline further.

A number of factors play a role in this process, including population pressures and deteriorating environmental conditions. Equally important, however, is the fact that agricultural growth takes place mainly in areas devoted to commercial export crops, which tend to use modern technology and inputs increasingly, often displacing the use of labor, except on an unstable seasonal basis. Despite numerous instances of successful small-scale commercial farming of non-traditional export crops, in general agricultural development tends to become polarized, leading to the concentration of land, resources and incomes in relatively fewer hands. Rural poverty is certainly the major unresolved economic and social problem in the developing countries, involving hundreds of millions of human beings. While local level projects to alleviate poverty abound, such efforts are rarely able to address its root causes, which are embedded in the more general tendencies of historical development and underdevelopment. To the extent that the world's poor are being denied their essential economic and social rights as a result of macro-economic policies that lead to the conditions described above, it may also be argued that they are particularly vulnerable to abuses of their civil and political rights.

Absolute and relative poverty on the one hand, and glaring economic disparities on the other, create conditions for social tensions and political conflict, particularly when, as is so often the case, inequality is accompanied by ethnic differences, discrimination and racism. In modern mass society, as traditional communal structures break down, people tend to mobilize for social and political action in emerging political party structures as well as in social movements of various kinds. But many societies have not developed the institutional mechanisms through which social mobilization can be successfully achieved without major conflicts. The existing power structure is often resilient to change, and in countries without proven democratic traditions, protest movements that channel popular demands may shatter against rigid authoritarian regimes. For many decades, after the period decolonization, one party states ruled in the then so-called Third World countries, and civil organizations hardly had a chance to make their voices heard. In Latin America, civil society re-emerged during the nineteen eighties, during what some authors call a 'third wave' of democratization, when a number of countries were able to re-establish civilian regimes after years of authoritarian and highly repressive military rule. Under these unstable conditions, the violation of human rights had been more the rule than the exception.

4.2 Democratic Consolidation?

Democratic political structures are evidently a primary condition for the effective enjoyment of human rights. But democratic politics cannot be reduced only to a transparent and credible electoral system (which is itself difficult to achieve in countries with highly unequal economic and social structures), nor to the periodic alternation of parties in power. Real democracy requires much more: it needs legitimacy, accountability, an independent judiciary system, and effective mechanisms for the veritable democratic participation of people from all walks of life in the affairs of governance at every level and in the daily mechanisms of decision-making which affect their livelihoods.

In numerous countries where poverty and socioeconomic inequalities persist and where the promise of development has not been fulfilled, observers note that people are disillusioned with the existing political party systems and with recurrent elections which do not actually provide effective alternatives. This disappointment has led to high levels of electoral abstentions or else to massive protest votes against traditional parties in favor of unproven if articulate 'strongmen' (such as the electoral victory of Hugo Chávez—a military officer who had attempted a coup in 1992—in Venezuela in 1998), who then may turn into a new kind of authoritarian figure (such as president Fujimori in Peru, who won a democratic election and then dismissed the country's congress and had himself re-elected). In an atmosphere of corruption, political cronyism, electoral fraud, and blatant disregard for the needs and aspirations of the majority population, it is little wonder that human rights are short-changed. And this occurs regardless of human rights legislation and public institutions for the protection of human rights (Ombudsman). In fact, human rights legislation is fairly widespread in Latin America, but its application and implementation are wanting. Also, most Latin American states have now set up public human rights commissions or offices, but there is still a long way to go until we may speak of an effective 'human rights culture' in Latin America.

4.3 Indigenous Peoples

None have been more negatively affected by such processes than the indigenous and tribal peoples, numbering around three hundred million, mainly in Asia and Latin America, who probably represent the weakest and most vulnerable segment of the world's rural populations. The idyllic and quasi-paradisiac image that popular literature presents of these peoples is far from reality. In general, they have been increasingly ravaged by progress and development. They are victims of genocide and ethnocide, their environment has been destroyed, their territories have been invaded by outsiders, their lands and livelihoods taken, their languages and cultures suppressed and discriminated against.

While some national governments take measures to improve their condition, and international organizations have become increasingly active on their behalf, in general indigenous and tribal peoples have truly become 'victims of development'. Just as millions of non-indigenous rural folk, the indigenous have also increasingly taken to the migratory labor circuits in the 'factories in the field', the megaprojects (road building, hydroelectric dams, oil fields etc.), or else a one-way ticket into the urban shantytowns and slums, where researchers have studied the phenomenon of 'urban marginality' since the nineteen sixties. The problem of 'marginality' turned out to be much more complex and diversified than was indicated simply by demographic growth in squalid settlements on the fringes of urban centers. Marginality came to mean not only lack of access to social and urban services, reflected in low standards of living, but also, and perhaps principally, lack of integration into the formal labor market and the structured economy. Indeed, levels of open unemployment and underemployment were persistently high among the so-called marginal population. In fact, marginality came to be identified with underemployment. And while it was hoped that the acquisition of modern skills through formal education and vocational training would pave the way for the full integration of the marginal population into the modern sector, it turned out rather that economic growth patterns were not providing the hoped-for employment opportunities on a sufficient scale to offset the expanding phenomenon of urban marginality. Formerly simply marginalized, today's poor can be seen as increasingly excluded from development. Whereas in the industrialized countries the 'excluded' sectors constitute distinct minorities (the undereducated, the drop-outs, the unskilled migrants, the prematurely laid off), in the developing countries many millions continue to be structurally excluded from the formal economy and the high-productivity sectors because the latter are simply not designed to absorb the growing pool of labor that has been formed by the breakdown of the traditional village economy and continuing high birth rates. Thus, today's 'emergent economies', as they are euphemistically called, are characterized, among other elements, by permanent under- and unemployment, the vast extension of the so-called informal economy, the inability of local ecosystems to deal with rapid technological changes, the subordination of national development needs to the demands of the export-driven, profit oriented but hardly employment-generating high growth sectors, as well as by their inability to provide for the satisfaction of the basic human needs of their majority populations. This holds for most countries of Asia, Africa and Latin America where the greater part of the world's population strives to make a living.

4.4 The Struggle for Indigenous Human Rights

The indigenous population of Latin America is estimated at around 40 million, which means roughly 10 % of the region's total inhabitants. This population is distributed quite unevenly, being concentrated in the Andean countries as well as in Mexico and Central America. In two states—Bolivia and Guatemala—the indigenous are the majority of the national population, while in some others (Ecuador and Peru) they make up more than one fourth. Mexico has the largest number of Indians, and while they are strongly concentrated in the country's central area and the southeast, they only represent around 12 % of the country's total. Elsewhere, as in Brazil and Argentina they make up only a small percentage. (Peyser and Chackiel 1994) Estimates also vary about the number of indigenous peoples, but taking mainly linguistic criteria (whether they speak a distinct indigenous language), we are referring to around 400 different groups, who are in turn divided into many thousands of local communities. Some native speakers (such as the Maya and the Quechua number in the millions) whereas others (such as numerous Amazon tribes) are on the verge of extinction.

Ever since the Europeans first came to the shores of the American continent to conquer and settle, indigenous peoples have suffered discrimination, exploitation and racism. During 300 years of colonial domination, (from the voyages of Columbus to the beginning of the nineteenth century when most states in Latin America successfully established their political independence from Spain), indigenous societies were subjected to the worst forms of oppression and exploitation. Much of the colonial wealth of Europe was based on the use of servile Indian labor in the mines and in the fields. Whilst the term genocide had not yet been coined, indigenous cultures were destroyed or subordinated to the dominant Iberian Catholic mold, and frequently the victims of the widespread physical destruction of indigenous societies which accompanied the expansion of the colonial economy in North and South America.

Still, the Spanish colonial empire adopted certain measures for the protection of its native vassals. Decimated as a result of military conquest, ecological destruction, forced labor and the introduction of deadly diseases brought by the colonists against whom the Indians had no defenses, the indigenous population decreased drastically in the century following upon the European invasion, only to begin recovering more than 200 years later. For the Indians, the first two centuries of colonial domination represented a demographic catastrophe, as their numbers plummeted.

The nineteenth century brought independence and a new legal and political system, controlled by the small, land-holding ruling class known in the region as *la oligarquía.* The expansion of agrarian capitalism and the modernization of the economy did not bring Indians many benefits. On the contrary, numerous indigenous communities lost their lands and were forced into peonage on the large estates, particularly during the reforms of the mid-nineteenth century. They were excluded from full participation in the economic, social and political system in an unequal relationship that has at times been described as a 'caste system' in which the indigenous peoples occupied the lowest strata of the social pyramid.

Special legislation often placed indigenous populations at a disadvantage in relation to the rest of society, even when some laws were of a protective and tutelary nature. While formal citizenship to all nationals was granted in some countries after independence, in others Indians were treated as minors and as legally incompetent until very recently. This situation of inequality and relative disadvantage lasted well into the twentieth century, and only began to change towards the nineteen-fifties, when the traditional land-holding system began to fall apart and economic modernization affected even the most backward areas. (In Mexico, the process began earlier, after the Mexican revolution of 1910). But economic development during the twentieth century has been highly unequal in Latin America, and the benefits of economic growth were (and still are) concentrated at the upper end of the social and economic scale. Poverty and extreme poverty are widespread all over rural and urban Latin America, and the indigenous peoples are mainly concentrated in the lower levels. A World Bank report published in 1994 declares that the living conditions of the indigenous people are abysmal, and that their poverty is persistent and severe. Under these conditions, they are particularly exposed to various kinds of human rights violations (Psacharopoulos and Patrinos 1994).

Poverty and economic inequality explain much of the 'underdevelopment' and 'backwardness' of Latin America's Indians in relation to the rest of society, but a deeper problem is the racism and discrimination of which Indians have been the perennial victims within the social institutions of the wider society. To the extent that Latin America's population is increasingly mestizo, that is biologically mixed, discrimination and racism are not so much based on perceptions of biological superiority and inferiority, than on cultural distinctions. These result in turn from the prevailing dominant idea of the nation-state based on Western, European or Mediterranean values which ignore, deny or actually reject the non-Western indigenous components of the national cultures of Latin America.

Thus, indigenous peoples, qua indigenous cultures, with their own identities, traditions, customs, social organization and world-view never did find a place in the process of 'nation-building' that Latin America's ruling classes embarked upon after political independence. Moreover, the *indigenista* policies adopted by Latin American states in the nineteen-forties were designed to 'integrate' or 'assimilate' the Indians into the national mainstream. The hegemonic nationalist ideologies of the twentieth century strengthened the self-perception of the ruling groups as nations without Indians; or at best as mestizo nations which had somehow effected a synthesis between the original European and Indian roots of nationality (to which sometimes was reluctantly added the African element), but whose cultural identity was in fact to a great extent a deliberately constructed 'Western' identity.

To their economic backwardness (as defined simply in the fashionable 'developmental' language of the times) and social and cultural discrimination, must be added political exclusion, because despite enjoying formal citizenship, indigenous peoples as such have not had much of an opportunity to participate as Indians in the political life of their nations. They were expected to assimilate and in fact to disappear as culturally distinct entities. To achieve this objective was the purpose of the school system, religious missionary activities, and the various social policies which addressed the 'Indian problem'. In the nineteen-forties, a number of states came together to set up a coordinated *indigenista* policy designed to improve the conditions of the Indian communities and to further their 'integration' into the national mainstream.

In the early fifties the International Labor Organization published a report on the living conditions of indigenous populations and in 1957 it adopted Convention 107 for the protection of indigenous and tribal peoples in independent countries. In Latin America, the ILO launched an ambitious 'Andean Project' in several countries, designed to help the development and assimilation of indigenous communities in several countries through an integrated approach. By the middle eighties, however, the limitations of the assimilationist approach became apparent, and in 1989 the ILO general conference adopted Convention 169 which has a more 'rights-oriented' language than the earlier convention it is meant to replace.

In the nineteen-seventies, the United Nations Human Rights Commission prepared a report on the situation of indigenous populations, and in the early eighties it set up a Working Group on this issue. One of the results of this activity has been the drafting of a *Universal Declaration of Indigenous Rights*, which is scheduled to be adopted by the United Nations General Assembly, within the framework of the International Decade of Indigenous Peoples (1995–2004). A similar process is now underway at the regional level in the Organization of American States. But both these documents have run into trouble as some government delegations express their reservations concerning the rights of indigenous peoples.

4.5 The Emergence of Indigenous Actors

By the nineteen sixties a number of emerging indigenous organizations began to mobilize, lobby and pressure national governments to change their policies and to take indigenous concerns into account (Stavenhagen 1998a). In some countries, indigenous peoples became involved in violent and revolutionary struggles or civil wars. Nicaragua's Sandinista government had to negotiate an autonomy agreement with rebellious Miskito Indians, while a 30 year long conflict in Guatemala in which Indians were both victims and participants ended in a peace agreement in 1996. Colombia's long standing civil war has involved indigenous groups, and in southeastern Mexico, peace negotiations stalled between the government and the Zapatista indigenous uprising which began in 1994.

These changes led in several countries to constitutional and legislative reforms which for the first time in Latin America's legal history recognize the existence of distinct indigenous cultures and languages and the specific rights of indigenous peoples. In the forefront of these changes are Bolivia, Brazil, Colombia, Ecuador, Guatemala, Nicaragua, and Panama, but legal reform is taking place almost everywhere. Unfortunately, legal reform by itself is not sufficient to change the situation of most indigenous peoples, even when it is strictly implemented, which is not at all the case in most of the countries which have formally at least recognized indigenous rights (Clavero 1994; Barié 1998).

The economic tendencies which have negative impacts on the conditions of life and survival of indigenous peoples have in fact accelerated over the last few decades under the policies of neoliberal globalization. Take agricultural modernization, for instance. The widespread introduction of commercial crops for export, based on the intensive use of costly inputs (mechanization, improved seeds, fertilizers, insecticides) tends to displace traditional subsistence agriculture, on which most indigenous communities depend for their survival. Increasing production costs and the need for economies of scale have favored the consolidation of larger agricultural units and agribusinesses, putting small subsistence farms at a disadvantage in highly competitive markets. Government agricultural policies, instead of helping small subsistence farmers overcome their handicaps, have in fact pushed the poorer peasants out of business and favored the concentration of larger agro-industrial enterprises or they have forced the small farmers to become increasingly dependent on, and therefore vulnerable to, the globalized agricultural economy. Many Indians are caught up in this maelstrom of change, and they become uprooted and displaced, virtual 'development refugees', increasing the ranks of migrant laborers both within as well as across national boundaries. Millions of indigenous peasants have thus become itinerant agricultural laborers and migrants to distant cities, sometimes in foreign countries.

Indian peoples who for centuries were the victims of those who coveted their lands and their resources, nowadays occupy the 'last frontier' in their countries, the areas that until recently had little appeal for the ruling classes and the transnational economic groups. This has now changed. From southern Chile to the Amazon jungle, from the highlands of the Andes to the forests of Central America, there is no longer any territory which is not of some interest to expanding world capitalism, either for its mineral wealth, oil deposits, pastures, tropical or hard-wood forests, medicinal plants and agricultural plantation potential, or its water resources for irrigation and the generation of electricity for the benefit of distant cities and industries.

Surviving indigenous peoples are the most recent victims of global capitalist development, and if these tendencies continue unabated, their chances of survival are becoming slimmer. Indigenous groups are not, of course, the only populations negatively affected by economic globalization, but not only is the physical survival and well-being of their members at stake, (numerous are the examples of malnutrition, disease, prostitution, and criminal violence associated with the encounter between indigenous groups and the representatives of global capitalism), but also their very existence as distinct societies and cultures is seriously endangered.

As a result, indigenous organizations—and their defenders—are anxiously involved in promoting a world-wide agenda for the defense of indigenous peoples' rights before it is too late. Some of the principal issues on this agenda are the following:

4.5.1 The Right to Land and the Recognition of Their Own Territories

To the extent that indigenous communities in Latin America have been traditionally linked to possession of land as a basic productive resource, the loss of their lands to the large estates, the agro-commercial interests or state-sponsored economic or urban development projects has led to progressive loss of livelihood and chances for survival. Indian lands are usually collectively held, and the present trend towards privatization of what remains of these communal properties (the process actually started under liberal regimes in the nineteenth century), is undermining the already fragile ecological basis of the Indian communities. Mexico in the thirties, and Bolivia in the fifties, among other Latin American countries, initiated agrarian reforms to favor the small peasant farmers (most of them Indians), but by the nineteen eighties counter-reforms were under way, and the land-base of the Indian peasant villages has been deteriorating progressively.

Closely linked to the land problem is the territorial issue. Indigenous peoples have been historically rooted in specific locations, in their original homelands, which in some cases constitute well defined geographical areas. Many of the Indian organizations now demand the recognition and demarcation of these territories as a necessary step for their social, economic and cultural survival. The Kuna people of Panama and the Yanomami in northern Brazil have obtained constitutional protection of their territories. The Mapuche in southern Chile and the Miskitos of Nicaragua, among many others, have been in the forefront of these struggles in their countries. Yet as many observers have pointed out, constitutional strictures do not guarantee effective protection of Indian peoples within the framework of widespread judiciary corruption and political pressures on local and national administrations.

The Colombian constitution of 1991 recognizes the traditional homelands of a number of indigenous groups and assures them of legal protection; it took intensive lobbying by indigenous organizations to obtain this legal victory. In Mexico, however, the negotiations between the Zapatista rebel army and the national government are stalled because of the latter's unwillingness to recognize any indigenous territory not already foreseen in the present political constitution of the country.

Convention 169 of the International Labour Organization, adopted in 1989, calls upon States to respect indigenous lands and territories, and proclaims the right of indigenous peoples to control their natural resources. This is a most important right, because many of the current conflicts over land and territory relate to the possession, control, exploitation and use of such resources. In a number of countries it is the State which keeps for itself the right to these assets, and in numerous instances multinational corporations are asserting their own economic interests over them, unleashing complicated conflicts over ownership and use-rights with indigenous communities, in which even multilateral agencies such as the Inter-American Development Bank and the World Bank become involved.

4.5.2 The Right to Their Own Culture

The *Universal Declaration of Human Rights* and the two *International Human Rights Covenants* establish the right of every person to participate in the cultural life of the community. In Latin America, however, (as in many other countries) governments have long attempted to impose an artificial 'national' culture over the original cultures and societies of indigenous peoples. While this has begun to change over the last few years, in order for indigenous cultures to be able to survive the destructive effects of globalization and economic modernization, it will be necessary to carry out policies designed to protect and stimulate them in all their variety and richness. Thus, indigenous organizations struggle for the freedom to speak and be taught their own languages, and to use them in administrative matters and the courts, to practice their religions, to live by their local social institutions, to create their arts and handicrafts, and to express their world-views and ceremonial life. These human rights must perforce be recognized as collective rights within the framework of multicultural and plurilingual states. In some countries legislation to this effect has already been enacted, in others the issues are hotly debated at various levels (Stavenhagen 1998b).

4.5.3 The Right to Indigenous Legal Systems

Customary indigenous law is practiced widely among Indian communities in Latin America, but the national state only recognizes codified state law. This situation generates numerous tensions and can lead to serious human rights violations of indigenous populations. Local administrations have begun to take indigenous customs and mores into account when applying legal norms or distributing justice, but contradictions between the two levels of law still abound. Indigenous organizations demand the right to practice their own (unwritten) legal norms in certain spheres and to use the national legal system only when it is in their interest. Law schools and research centers have begun to show interest in indigenous law which was previously entirely ignored, and parliaments have considered enacting legislation that would permit the coexistence of different legal systems. However, the Latin American legal tradition is not favorably inclined towards legal pluralism and these issues need to be threshed out in the coming years (Clavero 1994).

4.5.4 The Right to Territorial Autonomy, Self-Determination and Political Representation

The collective territorial and cultural rights referred to above can only be fully implemented if indigenous peoples are allowed the free exercise of their right to self-determination, as set out in international human rights instruments. That is why indigenous organizations demand the right to be identified as distinct 'peoples', and not simply as amorphous 'populations'. ILO Convention 169 in fact speaks about indigenous peoples, but does not accord them the recognition that international law might suppose. Usually, Latin American governments are suspicious of the term 'peoples' with regard to the indigenous, precisely because of the international implications of the right to self-determination, which governments usually reserve for established states. In fact, the right of peoples to self-determination is understood by indigenous organizations mainly as the right to local and regional autonomy, and has never been interpreted as implying secession or separation from an existing state. Some Latin American states are more open towards these demands, whereas others, such as Mexico, are deeply suspicious of them.

Whereas the effects of economic globalization have in general been disastrous for indigenous peoples, the current crisis also opened up new perspectives for them. As national states become increasingly incapable of caring for the basic needs of their populations, particularly for the poorer strata who are in the process of becoming national majorities, and even as the trend towards privatization of land and resources undermines indigenous communities, the will to resist and to prevail generates a mobilizing effect that may lead to the greater empowerment of indigenous organizations. This leads in turn to new political relations in whom indigenous peoples may finally find the respect and recognition they have been denied for so long, and where the secular struggle for their human rights may at last be rewarded.

4.6 Conclusions

In this paper I have referred mainly to structural conditions of human rights as they affect the world's poor in a globalized economy, with particular reference to indigenous peoples in Latin America. I argue that the full enjoyment of human rights is problematic in situations of extreme poverty where great economic inequalities exist, such as are found today in many of the countries formerly referred to as the Third World. The social tensions generated by these conditions as well as other historical factors (post-colonial politics, ethnic differences among the populations etc.) make it difficult for democratic institutions to flourish. The lack of effective democracy favors an atmosphere of disregard to human rights.

Latin America's Indians are particularly vulnerable in this regard, because of their long standing marginalization and exclusion from full participation in the national societies. Moreover, the dominant concept of the nation-state in Latin America has effectively excluded Indians from the mainstream.

Currently, an emerging indigenous peoples' movement challenges the dominant ideas of neoliberal economic development and the hegemonic model of the nation-state. Indigenous organizations demand the recognition of their human rights, including group or collective rights, related to issues such as land and territory, cultural identity, social organization, customary law, autonomy and self-determination. In recent years, some Latin American countries have effected constitutional and legislative changes which may favor the implementation of indigenous rights. But legal instruments are not enough: changes must also take place in the institutions and mechanisms for their application, the judiciary must be reformed, legal and cultural pluralism must be recognized and respected, and democratic political processes must be strengthened. Within this framework, the conditions for a new kind of multicultural citizenship may arise. All of the above goes hand in hand with necessary changes in the prevailing model of globalized economic growth, which in general has affected indigenous livelihoods negatively, in the direction of sustainable and participatory development. Efforts to bolster indigenous rights at the international level (United Nations, Organization of American States) have so far produced limited results, with the important exception of Convention 169 of the International Labour Organization, ratified only by a small number of states, where it has, albeit, become national law. This has helped indigenous organizations in their struggles for the effective implementation of human rights.

Finally, let it be said that in socially and ethnically divided societies, such as Latin America, open and subtle forms of discrimination exist which will only disappear gradually after lengthy efforts at popular education and the creation of truly democratic civic cultures. Without such changes human rights will continue to be an aspiration for the future, rather than an everyday reality.

References

Barié Kolb, Cletus Gregor, 1998: *Los derechos indígenas en las constituciones latinoamericanas contemporáneas* (unpublished thesis).

Clavero, Bartolomé, 1994: *Derecho indígena y cultura constitucional en América* (México: Fondo de Cultura Económica).

CLCDS [Comisión Latinoamericana y del Caribe sobre el Desarrollo Social], 1995: *Informe de la Comisión Latinoamericana y del Caribe sobre el Desarrollo Social* (Santiago de Chile: BID, CEPAL, PNUD).

Peyser, Alexia and Juan Chackiel, 1994: "La población indígena en los censos de América Latina", in: *Estudios sociodemográficos de pueblos indígenas* (Santiago: CELADE, CIDOP, FNUAP, ICI): 27–48.

Psacharopoulos, George and Harry Anthony Patrinos (Eds.), 1994: *Indigenous People and Poverty in Latin America. An Empirical Analysis* (Washington, D.C.: The World Bank).

Stavenhagen, Rodolfo, 1998a: "Indigenous Peoples: Emerging International Actors", in: Crawford Young (Ed.): *Ethnic Diversity and Public Policy* (New York: St. Martin's Press).

Stavenhagen, Rodolfo, 1998b: "Cultural Rights: a Social Science Perspective", in Niec, Halina (Ed.): Cultural Rights and Wrongs (Paris: UNESCO).

United Nations Development Programme, 1993: *Human Development Report 1993* (New York: United Nations).

United Nations, 1995: *World Economic and Social Report 1995* (New York, United Nations).

UNRISD [United Nations Research Institute for Social Development], 1995: *States of Disarray. The Social Effects of Globalization* (Geneva: UNRISD).

Chapter 5
What Kind of Yarn? From Color Line to Multicolored Hammock: Reflections on Racism and Public Policy (2001)

Abstract In 2001 the *World Conference against Racism and Discrimination* took place in Durban, South Africa, and a number of side events on related subjects were organized at the time to provide an input to the discussions of the conference. Again UNRISD invited me to join a group of international scholars and activists to discuss issues related to racism and public policy, to which this chapter contributed.

At the turn of the twentieth century, W.E.B. du Bois the pre-eminent intellectual of the African-American people presciently foretold that this would be the century of the 'color line'. During the decades that followed, the world witnessed the rise and fall of Nazism and the Holocaust, the civil rights movement in the United States, the end of colonialism and apartheid, the emergence of indigenous peoples as political actors on the international scene, the renewal of racism in Europe and the horrendous spectacle of ethnic cleansings and genocides in Bosnia and Rwanda. And yet a century later, the 'color line' is still with us, separating peoples and cultures, dividing the powerful from the downtrodden, even as it binds some people together in tight ethnic communities but also ties up a lot of people in conceptual knots. So the color line turned out to be a string of many features and multiple uses; perhaps like a clothes-line on which we can hang out our dreams and dreads and dramas to dry. But the color line can also be seen as an interconnected net of multi-colored yarn, strung and woven together, yet each one fiercely singular. In my part of the world hammocks are made of multicolored yarn; if you attempt to lie on each separate strand, it will break, but if you stretch your hammock and relax on it, you can rest and dream and even make love. What kind of yarn, what kind of story makes up these multicolored hammocks?

There are, of course, many kinds of racism, diverse racisms—it is a monster with many faces. Nothing further from reality than the widespread idea that 'racial

This is an original text.

Texts and Protocols 4, DOI: 10.1007/978-3-642-34153-3_5,

prejudice', an irrational feeling of antipathy and rejection of some Other deemed inferior and not worthy of our respect and understanding, is a matter of individual choice at worst, and at best the result of ignorance and personal prejudice which can be overcome by logical arguments and well-intentioned educational projects. Not that prejudice and subjective attitudes of rejection do not exist; they do indeed, and they need to be dealt with, but they do not float freely in the abstract mind; they are implanted and cultivated by social and political conditions and circumstances which reflect the dynamics of complex group relationships. Unless we are able to come to grips with these issues the struggle against racism will turn out to be a bit like preaching against sin: it may allow us to take the moral high-ground but how effective will it be?

If we look back upon the last fifty-odd years since the founding of the United Nations, we see that thinking about racism has undergone some important changes.[1] During the first phase, racism was identified mainly with the legacy of Nazi ideology—the murderous, genocidal hatred instilled in the German nation against all the so-called inferior races, particularly the Jewish people, but also Gypsies, Africans, Slavs, homosexuals and others. Nazi racism was based on a carefully constructed pseudo-scientific ideology of racial purity and superiority, which has its roots in numerous strands of Western thought and found its way into the language of academic anthropology, biology, psychology and other disciplines. The Nazis promoted 'race science' or 'raciology' (*Rassenkunde*) in their universities to provide legitimacy for their perverted world-view. Today scientific racism no longer commands any academic recognition whatsoever, but can still be found under various guises in some scholarly institutions and publications.[2] The first activities of the UN in the struggle against racism related to eliminating this poisonous legacy from the post-war world, and the Universal Declaration of Human Rights of 1948 well expresses this concern.

The next phase relates to the struggle against colonialism and the fight of colonized peoples everywhere for freedom and national liberation. The struggle against apartheid belongs in this phase, though it took many more decades to achieve its objective, a free, plural and democratic South Africa. The nineteen fifties and sixties saw numerous former colonies achieve independence and statehood, and also witnessed the civil rights movement in the United States. Colonial racism was formally abolished, but its effects still linger on in many parts of the world. The UN proclaimed the right to self-determination in the *Declaration on the Granting of Independence to Colonial Countries and Peoples* of 1960, later incorporated as Article 1 of the Human Rights Covenants adopted by the General Assembly in 1966.[3] Racism here was considered more than a set of individual

[1] Early work on race and racism was carried under the auspices of UNESCO. See UNESCO (1956) and Kuper (1975).

[2] Barkan (1992).

[3] "Article 2. Everyone is entitled to all the rights and freedoms set forth in this Declaration, without distinction of any kind, such as race, colour, sex, language, religion, political or other opinion, national or social origin, property, birth or other status."

rejectionist attitudes; rather it was seen as an expression of the unequal relationship between peoples of different stock in a given historical setting. Emphasis turned from individual attitudes and structured racist ideologies to the rights of peoples and the building of a new, more equitable international order. The rise to prominence of the Third World framed the background to a new scenario of international inequities, later to be accentuated by the process of economic globalization, which to many observers appears as a new form of global racism.

During the seventies and eighties racism re-emerged in a new guise, this time in the industrial heartlands of the North, involving mainly migrant laborers from the periphery, refugees and former colonial subjects. Incidents of racist violence, including riots, increased in the urban neighborhoods of Western Europe, whose principal victims were Africans, Asians, Muslims and Caribbeans. Racial discrimination was reported in the areas of education, housing, employment, health services and the criminal justice system, in which the youth of racial minorities have been particularly singled out through a process of 'criminalization'. Besides Blacks, Latinos have been prominent victims of racial profiling and discrimination in the United States.[4]

A number of states began to see racism not as a series of isolated incidents, but rather as a patterned and structured social problem, and soon government action and international attention were brought to bear on the topic. Massive transnational migration flows provoked widespread political debates about the perceived dangers of too many foreign migrants, the need for demographic 'balance', the control of borders and so forth. Latent racism became manifest once again, and politicians thrived by playing the 'immigrant-racial' card. The emergence and voter appeal of extreme right-wing political parties raised the issue to new levels. From Enoch Powell to Le Pen to Haider, the new right saw in foreign immigration—meaning racially distinct migrants from Third World countries or former colonies—the specter of an endangered national identity being swamped by alien hordes of inferior stock. Some states enacted anti-discrimination legislation and new immigration laws, others set up commissions to study racial issues, and the European Parliament prepared reports and passed resolutions on the topic. Racism in Europe had once more become an international issue of concern.

The nature of the debate was changing, however. Few people openly advocated racial discrimination of the phenotypical variety, and in the new global environment, the very concepts of race and racial relations were undergoing transformation. As immigrant communities mushroomed in the industrial states, perceived biological distinctions meshed with recognized cultural differences. In some countries, 'race relations' became a code word for relations between culturally differentiated communities. Human rights defenders were now no longer advocating just general equality (which seemed to many to be unattainable), but a new concept: the right to be different. States were expected to become less

[4] After the terrorist attack on the US in September 2001, Arabs have also become the target of racial profiling.

assimilationist and more pluralistic. Cultural differences were not to be abolished, but respected and celebrated. The always elusive melting-pot was to be replaced by a spicy multi-cultural salad bowl.

The debate now shifted to culture. The extreme right pounced on the concept of the right to be different and appropriated it. Indeed, they said, if everybody else has a right to be different, so do we: the authentic national element, the true bearers of national identity. And so, successively, the right to be different has become an argument for closing borders, forcing assimilation, eliminating bilingual education, excluding the 'undesirable', the 'unassimilable' from the truly national. Taken to its extreme, this argument leads to ethnic cleansing, the current face of genocide. Ethnicity has now replaced racialism, and ethnic discrimination is the new face of racism in today's globalized multicultural world.

In a more subtle vein ethnic discrimination finds intellectual support in liberal arguments concerning democracy and development. While it is no longer respectable to blame so-called inferior races for their own misfortunes, some academics, harking back to fashionable theories of the nineteen forties and fifties, have rediscovered culture as the real culprit of economic backwardness and authoritarian political regimes. Forget the legacy of slavery and colonialism and the functioning of the international capitalist system. It now turns out, we are told [by Harvard Professor Samuel Huntington and his colleagues], that the value systems of certain cultures and civilizations are favorable to progress and democracy as understood in the West, whereas other cultures (in Africa, the Arab world, Latin America and some Asian countries) contain value systems that are decidedly inimical to progress and democracy. Therefore, if there is to be any development here at all, these peoples will have to change their value systems, or 'we', meaning the West, will have to do it for them. Is there much difference in this approach from the 'civilizing' mission that colonialism attributed to itself a century or so ago?[5]

Colonialism—like racism—is a creature of many faces, and even though we are now said to live in a post-colonial era, and have developed post-colonial languages and discourses to account for this transition, a closer look at the contested spaces of those imagined communities we like to call nations, reveals patterns of domination and exploitation, often accompanied by multiple forms of racism, which we may refer to as internal colonialism. Indeed, the perennial victims of internal colonialism in many parts of the world have been the indigenous peoples, and their assertive emergence in recent decades expresses their accumulated hurts and frustrations, as well as their age-old aspirations and dreams. The rights of indigenous peoples are central to the latest developments in the international struggle against racism, having received increasing attention in an emerging field of international law, as documented in current United Nations covenants, declarations and resolutions.

---

[5] Harrison and Huntington (2000).

The plight of indigenous peoples, often thought to be mainly an issue in North and Latin America, is in fact a worldwide phenomenon. Whereas in the past the genocide of indigenous populations together with the slavery of Africans has most poignantly expressed the most revolting aspects of the legacy of colonialism—and let me add right away that neither the indigenous nor the Sub-Saharan Africans were the only victimized peoples, though perhaps the most widely known—in the current debates on human rights, democracy and development, the indigenous peoples around the world have established an agenda for the fulfillment of their human rights and the attainment of justice and well-being that includes the struggle against various forms of discrimination, exclusion, marginalization and racism [which characterizes their situation and mars their chances for decent and dignified conditions of life].[6]

As we contemplate the achievements of the last three decades of the combat against racism and look at the tasks ahead, and as we witness the current controversies and difficulties in finding common ground on which to join forces in the future, we need to recognize the challenges that the various forms of racism—the various racisms—present both at the analytical and theoretical levels as well as at that of policy and praxis. Indeed, one of the lessons learned during the twentieth century is that there is no easy fix on racism—the various features require different kinds of understanding and action. And this necessarily means a reassessment of the conceptual framework involving the usage of the term race.

There is now widespread consensus that the concept of race—its general usage to the contrary notwithstanding—is of little or no scientific value and does not rest on any hard facts. Scientists talk of human populations that are more or less genetically related. [In fact we as humans share most of our genes with a number of primates and other animals, and almost all of our genes with other human beings]. The genetic make-up which may make some of us different from other humans—and which has often been used to justify the use of the term 'race'—is absolutely minimal compared to the total human genome. The small numbers of physical traits that biologists in the nineteenth century used to identify the world's so-called major races are of no particular relevance to human social behavior. The concept of race became important as an instrument in the construction of explanations concerning the perceived differences between human populations that fascinated European travelers ever since the age of discovery, and later became an essential ingredient in the elaboration of theories designed to support ideas relating to the purported superiority of some peoples over others. It was perhaps inevitable that the principal use of the term 'race' served to justify the domination of one human group over the racially stigmatized 'Others'. And just as inevitably, the concept of race has become a weapon in the struggle for liberation, dignity and human rights by those so stigmatized. Race, then, became a socially, culturally and politically constructed signifier in a contextually determined system of 'race relations'. Under what circumstances and how do social relations become

[6] ICIHI (1987), Anaya (1996).

'radicalized'? Much research has been carried out on these questions, and some of it will be discussed at this Seminar.

As recent studies in Europe have shown, racism as a social phenomenon—not only an individual attitude—feeds on radicalized interpersonal relations and weaves the constructed concept of 'race' into its various ideological expressions. This is not only a Western phenomenon because it occurs wherever ethnically distinct peoples encounter each other in a patterned system of unequal and asymmetrical relationships. By ethnically distinct, I mean of course not only physically visible features but cultural group identities as well. Nor should we forget the phenomenon of counter-racism and reverse discrimination that we encounter in post-colonial societies and which complicate the tasks of building pluralistic democratic polities.[7]

To deal with these issues in more than a perfunctory manner it is useful to examine the various levels at which racism is shaped, expressed and experienced. The media and public opinion often tend to reduce racial discrimination to its subjective interpersonal expressions. To concentrate on individual racial attitudes and prejudice is to look at motivations, beliefs, stereotypes and values that may lead to discriminatory behavior patterns which sometimes include racist violence. Such prejudice involves complex mind-sets that under different circumstances may produce ethnocentrism, xenophobia, religious intolerance, and other attitudes often associated with insecurity, low self-esteem, suspicion, paranoia and a penchant for authoritarianism.

While such psychological factors may be present in many instances of inter-racial and inter-ethnic behavioral patterns at the personal level, it will not do, as some would have it, to reduce this level of racism to some sort of psycho-social dysfunction or label it as simply irrational. However, having been described too often as a malady to be cured, we should not take such approaches to racism lightly. Efforts to eradicate racist stereotypes, prejudices and attitudes must be continued at all levels, from personal counseling to group awareness-building procedures to educational activities to media campaigns. Much can be achieved at this level but only if the other levels of racism and discrimination are considered as well.

Institutional racism is surely at the present time the most widely debated expression of racial discrimination, xenophobia and intolerance. It refers to institutional practices that tend to place the victimized group in continuous disadvantage with respect to a majority or dominant group in society in a number of areas such as education, employment, career opportunities, housing, health care and other social services or societal goods or benefits that are thus unequally distributed along racial and/or ethnic lines. Institutional racism may not be the result of any personal racist motivation by people in positions of power, but it clearly affects the outcomes: biased recruitment patterns in jobs, unequal access to health care, limited career opportunities, lower quality of education and delivery of other social services, ghettoization, and multiple other forms of segregation and

---

[7] See, for example, Blackstone et al. (1998).

exclusion. Whether it is Blacks and Latinos in the United States, Caribbean youth in the United Kingdom, Arabs and Africans in France, Turks in Germany, indigenous peoples from Argentina to Alaska to Australia, Burakumin in Japan, Dalits in India, Berbers in North Africa, the patterns of institutional racism tend to be similar the world over. They are frequently not even formally considered as racist, and may appear under the mask of social and economic disadvantages simply suffered by lower income sectors. This is the debate surrounding the issue of descent-based and work-based discrimination among Asia's untouchable castes.

Disaggregated national development statistics often present such an undifferentiated picture: the underprivileged are a lower percentile, which is all. The United Nations Development Report has begun to correct this unsatisfactory vision: the breakdown of national welfare and development indicators by ethnic and racial groups shows a vastly different side of the story, as any serious research on these issues also demonstrates.[8]

Very often the underprivileged groups (a loaded term, no doubt) are blamed for their own misfortunes: it is said that they are not ambitious enough, their family structures are dysfunctional, their cultural values are traditional, their motivations are misplaced, and their world outlook is inadequate. In the standard language of our times, the victims are at fault, not the system. We hear arguments that in a liberal democratic state racial and ethnic discrimination is an aberration and should not happen at all. If discrimination is now outlawed in most democratic states and the legal system establishes that every one is equal under the law, then surely, if it still occurs, as is often claimed, the victims are partly to blame. There is much debate in the West about whether 'equality of opportunity' should lead to 'equality of outcomes'. Some recent scholarship holds that development actually means more freedom of choice based on enhanced capabilities of the individual. A just society would allow all individuals equal opportunity to increase their capabilities, and therefore overcome traditional inequalities.[9] But what if inequalities are persistent over decades and centuries and related to community, religion, ethnicity, culture or racial distinctions and to a history of oppression and exploitation?[10] Equality of opportunity, as we know, is not universally enjoyed, not even when the legal system is open and basically fair. Too often racial and ethnic discrimination occurs in the functioning of legal institutions, in the realm of the administration of justice, and particularly in the criminal justice system.

As considerable amounts of research have shown over the last few decades, in racially and ethnically divided societies social, economic and political institutions can be highly biased and produce slanted results. Much attention has been paid to the identification and implementation of public policies designed to eliminate racial and ethnic—and also gender—discrimination in institutional performance of

[8] The Human Development Report is published annually by the United Nations Development Program.

[9] Sen (1999).

[10] Tilly (1998).

all kinds. If racism is considered to be a social problem, as in so many immigrant countries, then social services—such as health, education and housing principally-must be made to work equally well for all citizens and denizens. Perhaps special efforts will have to be made to improve the situation for the most disadvantaged, the socially excluded, the especially vulnerable (such as the undocumented migrant worker or refugee, the untouchable caste, the marginalized indigenous community, the inner-city ghetto youth etc.).

This is what 'equal opportunity' in employment or any number of affirmative action policies, preferential politics, special development efforts and other public policy measures have set out to do. How effective are they? The report card comes up with mixed grades. In some cases affirmative action has indeed produced important results, in others it has become entangled in bureaucratic quagmires. Frequently it is accused for being a code word for reverse discrimination, and in the U.S., for example, it is now being dismantled by the courts.[11]

The argument against affirmative action goes somewhat like this: it creates more dependency among the underprivileged minority vis-à-vis the state, it does not benefit the poorest majority but rather the minority elites, it lowers the quality of education, it denies the principle of individual equality, it runs counter the basic values of a free, individualistic meritocracy, it violates human rights.... To hear some people say it, affirmative action turns out to be more discriminatory than the historical discrimination it was intended to redress. Minorities who benefit from affirmative action are rightly worried that the gains obtained over the past decades will quickly be eroded and that the indicators revealing unequal access by disadvantaged minorities to social benefits will rise once more.

The term affirmative action covers a variety of possible policy measures. These may not work under all circumstances, they can surely be improved upon, but they may be hugely successful in other cases. Certainly affirmative action policies must be used—and intensively so—to redress historical injustices that ethnic and racial groups have suffered, including colonized majorities as in South Africa. The reversal of affirmative action in the United States and elsewhere is not a policy that helps combat racism, but rather tends to entrench it.

In other situations, such as that of indigenous peoples in Latin America and elsewhere, compensatory measures for age-old discrimination take on other forms. Here the issue is not so much the possibility of individuals obtaining better access to the collective goods of society through existing institutions, but rather the design of institutions that will improve the life-chances and levels of welfare of disadvantaged communities and collectivities. In many parts of the world indigenous peoples have been deprived over the years of their homelands, their land and resources and the major requirements for their subsistence as peasant societies or as hunters and gatherers or pastoral nomads in fragile ecosystems. Here legal and institutional remedies to perennial discrimination include the restitution, demarcation and protection of traditional territories and homelands, agrarian reforms,

[11] Appelt and Jarosch (2000), Curry (1996).

investments in infrastructure, local and regional development projects of all kinds. In some countries special government agencies for indigenous affairs deal with these issues, but they are often accused of being too bureaucratic and paternalistic, when not downright authoritarian.[12]

Increasingly, ethnic minorities and indigenous peoples are demanding some sort of regional, political or cultural autonomy. While there are many different forms of autonomic relations between a state and existing minorities, these are complex issues that often involve rethinking the traditional nationalistic concept of territorial sovereignty, and not many states are willing to relinquish what they consider an essential element of their power and legitimacy. Indigenous peoples, for example, have insisted in the United Nations—and at the *World Conference Against Racism* held in South Africa in September 2001)—that they must be recognized qua peoples and that their right to self-determination (including autonomy) be respected. Some governments disagree on this issue, and that is why the draft UN Declaration on Indigenous Rights has been stalled in a working group of the Commission on Human Rights. Indigenous organizations have interpreted this resistance as yet another form of racism.[13] (Since this was written the Declaration has been adopted by the UN General Assembly).

Reparations and compensation for past injustices are the most recent issues on the negotiating table. Can centuries of genocide, ethnocide, colonial exploitation, slavery, debt peonage and oppression ever be repaid? There should surely be no question about the justice of such claims in principle, but putting them into practice is another matter. Who will be the direct beneficiaries of these claims, who should be deemed directly responsible for them, how would the process of adjudication work? Is any existing national or international legal system capable of handling such a process? What might the time limits, if any, be? A point of reference has been the restitution accorded by some European states to the survivors and the heirs of victims of the Holocaust or the claims of 'Comfort Women' in Japanese occupied territory during the Second World War. Can equally clear claims be made in the name of the enslaved or exterminated peoples of Africa and the Americas? Who will pay for the damages wrought upon the people of South Africa by decades of apartheid? Besides punishing some of the culprits under recent statutes in international criminal law, how can the horrors and the pain of contemporary genocides and ethnic cleansings in the Balkans and the African Lakes region ever be compensated for, not to speak of the politicides in Cambodia and the Soviet Union in earlier years, among others? These are some of the more recent challenges facing human rights policies and institutions in the world.

To go yet a step further, perennial racism and ethnic discrimination are deeply entrenched in the functioning of social institutions because differences, inequalities and hierarchies among peoples are firmly rooted in the overall dynamics of the

[12] Daes (2001).

[13] Statement by Menchu Tum at the World Conference Against Racism, Durban, South Africa, September 2001.

world system: economically, politically, socially and culturally, to use a simple classification. There is no doubt that personal identification, a sense of belonging, collective identity, ethnocentrism, group bonding and similar phenomena condition the relationship between the us and the other, between the we and the they. But whether, if and when these identifications become competitive, confrontational or cooperative will depend on the wider framework, and in today's world this means not only power relationships within the nation-state structure but also the global economy and its multiple implications.

We need not belabor the point that the global economy not only draws peoples closer together in a certain sense, but also generates new differences and inequalities. As Seattle and Genova have signaled so clearly there are new global postcolonial mechanisms of inclusion and exclusion afoot, which become particularly relevant during times of economic recession and retrenchment as is now the case. But regardless of cyclical phenomena, the globalization process itself continuously generates poverty, marginalization and exclusion, even as it tends to raise productivity, output and profits. At a safe historical distance, we can now recognize that Mr. Marx's prognostications were not so far off the mark after all.

The issue that brings us together here is the realization that the polarizing mechanisms of globalization have racial, ethnic and cultural implications. Far from being random phenomena, inclusion and exclusion are linked to the historically generated processes of ethnic and racial construction and differentiation, and the globally excluded, the persistently poor, the hungry, the sick (over half the world's population by United Nations estimates), are also the victims of discrimination on ethnic, racial and cultural grounds. Is not the poverty in the South amidst a world of plenty in the North (and these differential labels can coexist within a national space) a form of racism? Is not the destruction of viable and vibrant local communities and ecosystems due to the needs of capitalist accumulation a particularly severe form of discrimination? Is not the creation of fortresses of prosperity surrounded by worlds of misery and despair an extreme instance of intolerance and exclusion? Indeed, structural racism is the overall framework on which other expressions of racist and ethnic discrimination hang.

Is there a way out? Perhaps if we did not think so, we would not be gathered here today. Let me just say that I do not believe—as should be clear from the preceding remarks—that we can act against racism and racial discrimination as if they were isolated, self-contained phenomena. Public policies regarding racisms have tended to be reactive rather than proactive, remedial rather than preventive. In Western Europe, for example, state responses to racist violence have waxed and waned according to what some observers have identified as repeated waves of violence related to factors such as immigration and the political fortunes of extreme right wing parties at election time. Affirmative action policies, particularly in the educational field, may become popular when the social mobilization of discriminated minorities rises, and will be discarded or diminished at low levels of mobilization. Legislation on indigenous rights was introduced in Latin America in the eighties and nineties after the political emergence of new indigenous movements. Unless the pressure is kept up, such legislation will have little effect on the daily conditions of life of millions of indigenous. While formal apartheid has been abolished, the new global apartheid continues to affect the life-chances of millions of people the world over.[14]

As economic and social transactions between distinct communities and groups continue to be 'radicalized' in so many societies, the concept of race becomes socially relevant and racism is to seen as part of a system of power relations between radicalized actors, including not only individuals, but also institutions, the state and the global economy.

Blaming the 'system' in the abstract, however, is not a very constructive way of dealing with the issues; it leads to the old rather ineffective approach of saying 'we cannot do anything unless the system changes', but who will change the system and how? We've been through this before. As far as racism and racial discrimination is concerned, the locus of action has usually been the nation-state or even lower-level units. While global approaches are necessary—and the World Conference against Racism is an example of this—national and local level policies continue to be essential. Here we encounter a number of alternative approaches.

At the basic level of individual human rights, the struggle for equality has been a driving force throughout history. Wherever individual members of a discriminated group are disadvantaged through unequal treatment before the law or unequal access to opportunity and services of all kinds, or are politically, socially and culturally excluded from effective participation in violation of the basic principles of international human rights law, then any and all measures designed to overcome such disadvantages must be pursued. Experience shows, however, that simply removing legal barriers and proclaiming formal equality is never enough.

To remedy this situation, numerous countries have adopted some form of affirmative action or preferential treatment for members of discriminated groups and such measures have proven to be fairly successful as far as they go, though they have also led to counter-measures and challenges from groups who fear for their privileges.

The thrust for equality does little to address major group differences between ethnic and racial communities however, a question that has bedeviled debates on the relationship between the state and such collectivities, particularly when the latter are clearly in a subordinate position. States may face these issues by adopting various long-term strategies:

- Segregation of the subordinate groups. The patent failure of this policy is clear in the history of race relations in the United States and South Africa.
- Assimilation of the subordinate groups into the dominant society involving their disappearance as distinct cultural or ethnic peoples, a policy that has been implemented the world over in such disparate settings as indigenous Latin America, Berbers in North Africa, Moluccans in Indonesia and Kurds in western Asia, among others. Observers have described cases of forced assimilation as

[14] Witte (1996), Stavenhagen (1998).

forms of cultural genocide or ethnocide. This policy may be successful at times but usually entails a high social cost and major violations of human rights.

- A somewhat different situation prevails in countries that have important contingents of immigrants, where assimilation is said to be not only in the 'national interest' but also in the best interests of the immigrant groups themselves. While assimilation is often presented as a solution to the tensions and confrontations that accompany the settlement of immigrant groups, including refugees and asylum-seekers, it can also be said that the policy itself generates tensions and confrontations.
- A softer position—to label it some way—involves integration of subordinate groups into the dominant society, which really means incorporation into the dominant model of the nation state (which is often an ethnocratic state), at the same time respecting certain features of their collective identities (maybe the use of language, freedom of religion, local forms of social organization etc.).

Much more contentious is the recent emphasis on multiculturalism. As a consequence of what some have called an 'ethnic revival', partly resulting from the universalisation of the discourse of human rights, partly as an answer to the weakening of the nation-state, in part as a substitute for the decline of overarching social and political ideologies, the recognition and celebration of diversity leads to a new awareness concerning the role of culture in shaping social communities and conditioning individual behavior. Some countries–such as Canada–have adopted active policies of multiculturalism meaning the official recognition of numerous communal identities (Native Americans, linguistic communities, immigrant minorities) which has led to the adaptation of the legal system to the requirements of culturally differentiated collectivities within the federal state structure. Thus it represents a respect for the collective rights of ethnic groups, particularly those that for historical reasons occupy subordinate positions in the wider society. Within the framework of a renewed debate on the meaning of citizenship, the notion of multicultural citizenship has taken hold of public discourse and is being considered in many different contexts. To the extent that the denial of cultural identities is a form of racism, multicultural citizenship rights may be considered as an effective way of combating racism, discrimination and exclusion at the societal level. To be sure, it requires rethinking the idea of the nation-state itself. This is indeed the challenge facing a number of states in the former Soviet Union and Yugoslavia, where the long cherished idea of the multinational polity was violently destroyed during the nationalist conflicts of the nineties.[15]

To the extent that the reification of cultural differences may also become a form of exclusion and even racism, multiculturalism as an objective of public policy is receiving a good deal of criticism lately. It has other problems as well, such as the possibility of setting up conglomerates of legally fixed communal identities in a sort of corporate structure that may run counter the current world tendency towards democratic liberalism. Moreover, insofar as ethnic community structures may

[15] Kymlicka (1995).

impinge upon the individual human rights of their own members (for example, by demanding strict religious conformity or adherence to traditional marriage customs) legally sanctioned multiculturalism (in the form of autonomy, for example) may be at odds, according to some writers, with the idea of universal human rights.

Increasingly there is talk of interculturality rather than multiculturalism per se. This would not deny cultural diversity among groups but rather strengthen it through flexible structures of governance and socialization, within the context of state structures that are not culturally bound to any particular model of the 'nation-state'. How the idea of interculturality would play out in the fields of education, communication, social control, cultural creativity, administration of justice, political representation and so forth is still an open question. But the debate has begun.

Identity and identification, dignity and diversity, power and politics, rights and resources: these are some of the contested spaces in the struggle against discrimination and racism in our post-colonial, globalised world. How well we will be able to deal with them is one of the major challenges of our beginning century. Du Bois' color line has become a multi-colored hammock of interconnected yarn. Each thread contributes to the strength of the net. Let us make sure that the colors hold fast and the hammock does not break.

References

Anaya, S. James, 1996: *Indigenous Peoples in Interntional Law* (Oxford: Oxford University Press).

Appelt, Erna and Monika Jarosch (Eds.), 2000: *Combating Racial Discrimination. Affirmative Action as a Model for Europe* (Oxford: Berg).

Barkan, Elazar, 1992: The Retreat of Scientific Racism (Cambridge: Cambridge University Press).

Blackstone, Tessa, Bhikhu Parekh & Peter Sanders (Eds.), 1998: *Race Relations in Britain. A Developing Agenda* (London: Routledge).

Curry, George E. (Ed.), 1996: *The Affirmative Action Debate* (Redding, Mass.: Addison-Wesley).

Daes, Erika-Irene A., 2001: *Indigenous peoples and their relationship to land.* Final Working Paper presented by the Special Rapporteur. United Nations (E/CN.4/Sub.2/2001/21)

Harrison, Lawrence E. and Samuel P. Huntington (Eds.), 2000: *Culture Matters. How Values shape Human Progress* (New York: Basic Books).

ICIHI [Independent Commission on International Humanitarian Issues], 1987: *Indigenous Peoples. A Global Quest for Justice* (London: Zed Books).

Kuper, Leo (Ed.), 1975: *Race, Science and Society* (Paris: UNESCO).

Kymlicka, Will, 1995: *Multicultural Citizenship. A Liberal Theory of Minority Rights* (Oxford: Clarendon Press).

Sen, Amartya, 1999: *Development as Freedom* (New York: Alfred A. Knopf).

Stavenhagen, Rodolfo, 1998: "Indigenous Peoples: Emerging International Actors", in: Crawford Young (Ed.): *Ethnic Diversity and Public Policy* (UNRISD – Macmillan Press): 133-152.

Tilly, Charles, 1998: *Durable Inequality* (Berkeley: University of California Press).
UNESCO, 1956: *The Race Question in Modern Science* (Paris: Unesco Press)
Witte, Rob, 1996: *Racist Violence and the State* (London: Longman).

Chapter 6
The United Nations Special Rapporteur on the Rights of Indigenous Peoples (2012)

Abstract This is one of several texts that I have written on the mandate and role of the UN Special Rapporteur on the Rights of Indigenous Peoples, sometimes specifying my personal experience involved. It is intended to introduce the general public to the "special procedures" set up by the Human Rights Council of the UN, a human rights protection mechanism which is not too well known outside of this international body.

Increasingly concerned by the complexity of human rights issues in different parts of the world, the *United Nations Commission on Human Rights* (CHR) decided to establish in 1947 a *Sub-Commission on Prevention of Discrimination and Protection of Minorities*, composed of 25 independent human rights experts. Its mandate was "to undertake studies, particularly in the light of the Universal Declaration of Human Rights, and to make recommendations to the Commission concerning the prevention of discrimination of any kind relating to human rights and fundamental freedoms and the protection of racial, national, religious and linguistic minorities".

Among the many studies and reports produced by this body over the years, of particular interest for our topic is the S*tudy of the Problem of Discrimination against Indigenous Populations*, also known as the Martínez Cobo report. Although only the last chapter of this report, with concluding proposals and recommendations, was published by the UN in 1983, the full report is now available online. Paragraph 379 of the report proposes a provisional definition of Indigenous peoples which has continued to be widely used in the international environment, including the U*nited Nations Declaration on the Rights of*

This article was first published in: *Routledge Handbook on Indigenous Rights* (London: Routledge, 2012). The permission to republish this text was granted on 23 July 2012 by Ms. Cathy Hartley of Routledge.

Texts and Protocols 4, DOI: 10.1007/978-3-642-34153-3_6, © The Author(s) 2013

Indigenous Peoples (UNDRIP). This important study served for many years as a guide to the activities of the Sub-Commission's *Working Group on Indigenous Populations* (WGIP), established in 1982 and composed of five members, each one from a different geographical region.

This Working Group set an important precedent in UN practice by allowing extensive participation of indigenous representatives in its annual sessions in Geneva. Coming from many different countries, speaking in the name of numerous civil society, human rights and indigenous peoples' organizations, they became an articulate lobby for indigenous human rights in the corridors of the United Nations. The Working Group also produced a number of additional reports on specific concerns of indigenous peoples as, for instance, the *Study on Treaties, Agreements and other Constructive Arrangements between States and Indigenous Populations* in 1999, by Miguel Alfonso Martínez, and *Indigenous Peoples and their Relationship to Land*, by Erica Irene Daes, long-time chairperson of the Working Group.

Over the years, the WGIP, the Sub-Commission and the full Commission—transformed in 2006 into the *Human Rights Council* (HRC)—centered much of their activity on the preparation of a draft declaration on the rights of indigenous peoples, which finally became the UNDRIP, solemnly proclaimed by the UN General Assembly in 2007. Along the way, and responding to persistent demands of indigenous peoples, the Commission was also able to persuade the Economic and Social Council of the United Nations to approve the establishment of the UN Permanent Forum on Indigenous Issues, which meets annually in New York since 2002, and to create the mandate of a *Special Rapporteur on the Human Rights and Fundamental Freedoms of Indigenous People* (2001).

This mandate was set up within the framework of the Special Procedures adopted by the Human Rights Commission in the nineteen seventies when it became clear to its member states that certain human rights issues around the world could not be dealt with in routine fashion but required special attention by the Commission, which needed to be better informed about such issues before it could proceed to adopt specific resolutions that might lead to further action. Two kinds of mandates for special procedures were set up: country-specific mandates and thematic mandates. The mandate on indigenous rights is of the latter type, meaning that information must be gathered world-wide although references to particular countries are expected to be included in the special rapporteur's annual reports.

According to Resolution 2001/57 of 24 April 2001, the Commission on Human Rights decided to appoint, for a period of 3 years, a special rapporteur on the situation of human rights and fundamental freedoms of indigenous people with the following functions: (a) to gather, request, receive and exchange information and communications from all relevant sources, including Governments, indigenous peoples themselves and their communities and organizations, on violations of their human rights and fundamental freedoms; (b) to formulate recommendations and proposals on appropriate measures and activities to prevent and remedy violations of the human rights and fundamental freedoms of indigenous people; (c) to work in close relation with other special rapporteurs, special representatives, working groups and independent experts of the Commission on Human Rights and of the Sub-Commission on the Promotion and Protection of Human Rights.

At the initiative of GRULAC, the regional block of Latin American and Caribbean member states of the Commission, my name—along with other candidates—was presented to the Commission's chairman who, after consultation with other members of the 'Bureau' of the Commission, decided to appoint me as the first special rapporteur on the rights of indigenous peoples.

I took up my mandate in the summer 2001, when I arrived in Geneva to attend the annual session of the WGIP and to receive a briefing by the secretariat of the UN High Commissioner's Office for Human Rights who was, at the time, Mrs. Mary Robinson, former president of the Irish Republic. She expressed her great personal interest in the human rights of indigenous peoples, especially of indigenous women, and offered me the full support of the Office for my task. I also established a firm working relationship with the Indigenous Peoples and Minorities Section of the Office, then headed by Julian Burger. I soon realized that I was very fortunate to be able to count on the technical support and advice of this small Section in the Office, because not all Special Procedures (as we mandate holders were referred to in UN parlance) were able to receive this kind of support from the Office.

My first official mission that year was to attend the *World Conference against Racism, Racial Discrimination, Xenophobia and Related Intolerance* that took place in Durban, South Africa in September. This Conference marked the third United Nations Decade to combat racism and racial discrimination and concluded with an important final Declaration and Programme of Action, in which references were made to indigenous peoples in various contexts, although indigenous representatives who participated in the parallel conference of *non-governmental organizations* (NGOs) that drew hundreds of participants from all over the world, were not satisfied with the results.

Mostly, my time was spent in drafting my first report to the CHR to be presented at its fifty-eighth session the following spring. This first report (E/CN.4/2002/97) presented my general views on the situation of the human rights of indigenous peoples, based on earlier work done by the United Nations and its specialized agencies (which turned out to be much richer than I had imagined), and to propose a provisional work-plan for the subsequent years of my mandate.

I learned that it was routine for special procedures to seek approval of the Commission for their triennial work plan, although this did not preclude the possibility of modifying the program as the work progressed. A Commission resolution approving the report and program of the special rapporteur was required to legitimate the mandate and the incumbent, and served to encourage him/her to continue. Special rapporteurs are usually provided by the Office with an assistant who organized my country visits, helped gather information, prepared briefs, handled official correspondence and aided me with the drafting of my reports.

Contrary to ordinary belief, special rapporteurs do not get paid for their work at the UN and they are not considered employees or officials of the UN secretariat.

They are not representatives of their countries either, although they may have been supported or proposed by their government (as was my case). They are expected to work out of their own offices, institutions, organizations or homes, as the case may be, but be present at the UN for periodic meetings of various kinds and spend as much time on the mandate as they possibly can. In my case, I continued to maintain my tenured professorship at El Colegio de Mexico, but with a reduced teaching-load and the full support of the institution's authorities. My successor, Professor James Anaya of the University of Arizona, was able to build up more institutionalized support for his activity at that university. Not all special rapporteurs are that fortunate. Some are human rights lawyers or activists who must continue their regular professional activity for personal reasons and who may find their rapporteur's tasks too burdensome to maintain over an extended period of time.

To be sure, travel expenses for the special rapporteurs' multiple international activities are covered by the UN secretariat. Occasionally, these activities may also be funded by governments, foundations or civil society organizations when necessary, but they must in no way interfere with the SR's independence or judgment.

From the beginning of the exercise of the mandate we had to figure out a methodology for obtaining, classifying and analyzing the information and documentation relevant to the objectives determined in the Commission's resolution. In accordance with working guidelines and the activities carried out by other mandate-holders, there are three main lines of research available. The first are in loco country visits. This is the most significant way of obtaining information on human rights violations of indigenous peoples. From the beginning of my mandate I was literally besieged by invitations coming from indigenous organizations the world over. Country visits must be carefully planned beforehand and usually imply an intense agenda over a few days only, as each visit is generally limited to 10 or 12 days at most. An official visit to any country can only be arranged at the invitation of the government. If forthcoming, then the special rapporteur suggests an agenda and itinerary, based on previous knowledge of the situation in that country. This is then amended, accepted or rejected by the government. On several occasions I had to negotiate my agenda in the country carefully with government officials before reaching an agreement. Usually the SR proposes a visit after consultation with indigenous and human rights organizations in the country, whereas the government may be less interested in having the SR visit places of conflict and would prefer him/her to receive more briefings by government officials. Being in a country on a tight schedule and for a limited time only, an extra day with government officials means one less day in an indigenous community. To be sure there are also cases where officials would prefer to ignore the presence of the special rapporteur rather than host him.

Besides visiting government offices and receiving information from official sources, the rapporteur usually establishes contact with the diplomatic corps and international agencies that may be working in the country, such as UNDP, ILO and sometimes a local representative of the OHCHR. If there is one in the country, a visit to the human rights commission will be scheduled, as well as conversations with members of the judiciary, especially if there is judicial activity concerning the rights of indigenous communities. From my perspective, the most productive conversations were held during visits—however brief—to indigenous communities involved in conflicts or litigation over human rights issues, as well as meetings and consultations with civil society associations, human rights defenders and indigenous movements and organizations. During spare moments, which are few, the SR will be able hear complaints and receive further information from interested parties that request a meeting outside the official schedules. From my first country visit I requested meetings with academics and research institutions doing studies on the country's social problems and ethnic diversity whose experience usually turned out to be highly valuable for a better understanding of the local situation. And I always tried to give a talk or lecture on the mandate of the special rapporteur and on indigenous rights issues. From these encounters I obtained a wealth of information which was later carefully reviewed in order to incorporate the most significant findings in my country reports. Unfortunately, only a small fraction of the information obtained during these visits would find its way into the reports, because of formal reporting requirements and limited space provided to the mandate holders' presentations by the UN administration. This underutilization of important information and documentation was to be one of my many frustrations during the mandate.

A second major source of information is the documentation provided by governments, UN agencies and civil society and indigenous organizations at the request of the OHCHR and the SR. Every so often, we would send out letters and questionnaires requesting information on specific topics related to the thematic focus of the SR's forthcoming annual report. Much valuable information was obtained in this way, even though not all member states of the Commission answered such requests diligently.

A third source of information are the various 'communications' between the SR and specific governments on particular cases of alleged human rights violations involving indigenous individuals or communities. These communications are usually confidential until made public in the SR's annual report. Usually the exchanges of communications with governments over alleged violations of human rights stretch out over many months and only occasionally are there any documented satisfactory solutions to the complaints presented by indigenous people. More often, governments inform the SR that they are taking care of the problem and then nothing more is heard from them. Nevertheless, the SR needs to inform the Commission in his annual report about the state of communications with member states. Additional information comes to the attention of the SR from symposia and meetings organized by the OHCHR in support of the SR's thematic concerns. Thus, during my mandate, the Office in collaboration with national institutions organized a number of such meetings where specific human rights concerns of indigenous peoples were analyzed and discussed, as for example, education, legislation, and administration of justice.

The principal outcome of the special rapporteur's activities consisted of his periodic reports presented in writing to the CHR (now the, HRC) at its annual sessions in Geneva under (at the time) item 15 "Human Rights and Indigenous Issues". By the strict rules of the administration, the main body of the reports could not be longer than 11,000 words, but they could be enhanced, if required, by a number of addenda. In my case, the addenda included the reports on the official country missions I had undertaken during the previous year, a summary of the various communications between governments and the special rapporteur, and the report of the expert meeting organized around that year's thematic focus.

As I prepared my first systematic activities as special rapporteur in late 2001 I had to consider different options. The mandate as described in Resolution # 57 opened several possibilities, although it was clear enough that the CHR wanted me to look at the human rights violations of indigenous peoples. To be sure, the UN had done some prior work on the subject. There were the two decades of annual sessions of the WGIP, the famous but not widely known Martínez Cobo report, and ILO's Convention 169 adopted in 1989, which many considered—erroneously—to be mainly restricted to the traditional field of labor protection as understood by that specialized organization. Yet I knew that overall the diplomatic delegations present at the regular meetings of the Commission had little prior knowledge of (and perhaps not that much interest in) indigenous peoples and their rights. On the other hand, as soon as I had been appointed special rapporteur, indigenous and human rights organizations began to provide me with a constant stream of material involving precisely the human rights violations of their constituencies and expressing their expectation and hope that I would be their spokesman at the gathering of diplomats in the Commission. I realized how important this could become, especially because the Commission was much more concerned with the highly political human rights issues that emerged from a number of authoritarian or totalitarian states or that concerned the occupied territories of Palestine. Furthermore, the Commission had finally taken action after many decades on the rights of minorities by adopting in 1992 the Minority Rights Declaration. Numerous diplomats considered that indigenous populations would be well served by this Declaration and questioned their insistence on the need to produce an Indigenous Rights Declaration that was still being discussed in the Commission.

I decided that it would be useful to draw some of these loose ends together and to provide the Commission with some relevant information on the current situation of indigenous peoples and their human rights before exploring more specific topics related to indigenous human rights, such as detailed legal questions or concrete conflictive issues in particular countries. Consequently, my first report to the Commission, presented in March 2002, provides a panorama of the major human rights issues confronting indigenous peoples worldwide. These were grouped under various categories:

- Rights to land and territory and access to and control over natural resources. Based in information from different countries and other UN reports, I argued that "land, territory and resources together constitute an essential human rights issue for the survival of indigenous peoples" (para. 57)
- Education, cultural and language rights
- Multiculturalism
- Social organization, local government and customary law
- Poverty, levels of living and sustainable development. On these complex and interrelated issues I observed that "A new approach seems to be taking hold in international discourse: human-rights centered sustainable development, meaning that unless development can be shown to improve the livelihoods of people within the framework of the respect for human rights, it will not produce the desired results." (para. 83)
- Finally, political representation, autonomy and self-determination

Regarding numerous communications on specific cases of human rights violations, the report also mentions the problem of a 'protection gap' between existing human rights legislation and specific situations facing indigenous peoples as being of major significance and presenting a challenge to international mechanisms for the effective protection of human rights.

The problem of the human rights implications for indigenous peoples of major development projects was raised as the main focus of my second report to the HRC in 2003. I concluded, from an overview of much available information on this issue, that the principal human rights effects of these projects for indigenous peoples relate to loss of traditional territories and land, eviction, migration and eventual resettlement, depletion of resources necessary for physical and cultural survival, destruction and pollution of the traditional environment, social and community disorganization, long-term negative health and nutritional impact as well as, in some cases, harassment and violence against indigenous persons. The report recommended to governments that

> the human rights of indigenous peoples and communities must be considered of the utmost priority when development projects are undertaken in indigenous areas. Governments should take the human rights of indigenous peoples as a crucial factor when considering the objectives, costs and benefits of any development project in such areas, particularly when major private or public investments are intended. Potential long term economic, social and cultural effects of major development projects on the livelihood, identity, social organization and well-being of indigenous communities must be included in the assessment of their expected outcomes, and must be closely monitored on an ongoing basis. This would include health, nutrition, migrations and resettlement, changes in economic activities, levels of living, as well as cultural transformations and socio-psychological conditions, with special attention given to women and children.

Moreover, the issue of extractive resource development and human rights involves a relationship between indigenous peoples, governments and the private sector, which must be based on the full recognition of indigenous peoples' rights to their lands, territories and natural resources, which in turn implies the exercise of their right to self-determination. Sustainable development is essential for the survival and future of indigenous peoples, whose right to development means the right to determine their own pace of change, consistent with their own vision of development, including their right to say no. Free, prior, informed consent is essential for the human rights of indigenous peoples in relation to major development projects, and this should involve ensuring mutually acceptable benefit sharing and independent mechanisms for resolving disputes between the parties involved, including the private sector.

The impact of megaproject development on the human rights of indigenous communities has now become one of the most controversial issues pitting indigenous peoples against government authorities, private enterprise and international financial agencies. Indigenous Peoples are increasingly using legal strategies and judicial remedies as well as political lobbying and direct action to make their point, and are often suffering from government repression and the criminalization of their activities as a consequence. In some instances they have won reprieves or restitution in the courts, but in others the cards are stacked high against them. In my 2003 report I tried to make this situation clear to the CHR and made a number of recommendations to governments and development agencies. On these crucial issues of survival and wellbeing, indigenous peoples increasingly claim their right to free, prior and informed consent that has become Article 19 of UNDRIP since its approval in 2007.

In 2004 my report focused on the obstacles, gaps and challenges faced by indigenous peoples in the realm of administration of justice and the relevance of indigenous customary law in national legal systems. On the basis of research and numerous sources of information, the report indicated that

> Indigenous people tend to be overrepresented in the criminal justice system, are often denied due process and are frequently victims of violence and physical abuse. Indigenous women and children are particularly vulnerable in this respect. Numerous cases of criminalization of indigenous social and political protest activities have come to the attention of the Special Rapporteur. Language and cultural differences play their role in this pattern of discrimination, and they are not always sufficiently addressed by the State. Some countries have made progress in recognizing the specific needs of indigenous people in the field of justice and have adopted laws and institutions designed to protect their human rights. Indigenous customary law is being increasingly recognized by courts and lawmakers, as well as by public administration. Some countries are experimenting with alternative legal institutions and conflict resolution mechanisms, with encouraging results.

In several of the countries he visited, the Special Rapporteur has come across situations where there appears to be incompatibility between human rights legislation pertaining to indigenous peoples and other sectoral laws (such as legislation regarding the environment or the exploitation of natural resources, or the titling of private landholdings). When asked to rule on competitive claims on such issues, the courts may sometimes render judgments that protect the rights of indigenous communities, but just as often they may hand down rulings that are detrimental to these rights. The Special Rapporteur has always recommended that the rights of indigenous peoples as set out in national and international laws should have priority over any other interests and has called upon Governments to make efforts to adjust their legislations accordingly.

The widespread lack of access to the formal justice system due to ingrained direct or indirect discrimination against indigenous peoples is a major feature of the human rights protection gap. The overrepresentation of indigenous people in corrective institutions is often linked to overpolicing in areas where indigenous persons live and to the intense focus by enforcement bodies on indigenous activities, which leads to higher levels of arrests. Studies show that indigenous people are overrepresented in court, are charged with more offences than non-indigenous, and are more likely to be denied bail, spend less time with their lawyers and receive higher sentences when pleading guilty.

One of the more serious human rights protection deficiencies in recent years is the trend towards the use of laws and the justice system to penalize and criminalize social protest activities and legitimate demands made by indigenous organizations and movements in defense of their rights. Reports indicate that these tendencies appear in two guises: the application of emergency legislation such as anti-terrorist laws, and accusing social protestors of common misdemeanors (such as trespassing) to punish social protests.

An ominous trend in current affairs is that human rights abuses occur not only during states of emergency or in authoritarian non-democratic regimes, but also within the framework of the rule of law in open transparent societies, where legal institutions are designed to protect individuals from abuse and to provide any victim of alleged human rights violations with mechanisms for access to justice and due process. Rights abuses committed against indigenous people often happen in the context of collective action initiated to press the legitimate social claims of marginalized, socially excluded and discriminated against indigenous communities. Private vested interests and beleaguered authorities belonging to local power structures often use the law to dismantle such movements by penalizing prominent leaders either through the application of common criminal statutes and regulations or by invoking politically motivated anti-terrorist legislation. The Special Rapporteur strongly urges that legitimate social protest activity of indigenous communities not be so penalized by the arbitrary use of criminal legislation designed to punish crimes that endanger the stability of democratic societies. He urges States to use non-judicial means to solve social conflicts through dialogue, negotiation and consensus.

Through his study of the issue, and especially through his country missions, local visits and dialogue with leaders and individuals in the various communities around the world, the Special Rapporteur has found that a human rights protection gap with regard to indigenous peoples results from the operational deficiencies of the justice system, particularly in the area of criminal justice, and largely explains the widely reported lack of confidence of indigenous peoples in their national systems of administration of justice.

6.1 The Right to Education

This right figures prominently in United Nations human rights concerns and 2004 was also the last year of the UN's first international decade on human rights education. For this and other reasons it was appropriate to devote my 2005 report to the right of indigenous peoples to education. We organized an international seminar on the subject with UNESCO and I received much relevant material from governments and indigenous organizations. The report details a number of crucial human rights issues in this field.

The right to education is critical for millions of indigenous people throughout the world, not only as a means of extricating themselves from the exclusion and discrimination that have historically been their fate, but also for the enjoyment, maintenance and respect of their cultures, languages, traditions and knowledge. The systems of formal education historically provided by the State or religious or private groups have been a two-edged sword for indigenous peoples. One the one hand, they have often enabled indigenous children and youth to acquire knowledge and skills that will allow them to move ahead in life and connect with the broader world. On the other hand, formal education, especially when its programs, curricula and teaching methods come from other societies that are removed from indigenous cultures, has also been a means of forcibly changing and, in some cases, destroying indigenous cultures.

This situation has several aspects. First, there are the difficulties many indigenous people experience in gaining access to academic institutions. Secondly, many problems exist with regard to the institutionalization of educational services for indigenous people. Most problematic of all, however, is the fact that throughout much of history the fundamental goal of education has been to assimilate indigenous peoples in the dominant culture ('Western' or 'national', depending on the circumstances), a culture that is alien to them, with the consequent disappearance or, at best, marginalization of indigenous cultures within the education system. To a large extent, this is still the prevailing view in some countries' education systems, despite the existence of legislation that sets specific objectives in this area.

Aside from problems of discrimination in access to schooling, which are still widespread despite government efforts to eliminate them, an as yet unresolved human rights issue is that traditionally schooling for indigenous children had the purpose of assimilating them into the dominant society and separating them from their own cultures. An instance of this approach is provided by the story of the Residential Schools in Canada, which are now recognized as having done irreparable cultural damage to indigenous children in that country. The alternative approach to indigenous education in recent years has been to foster bilingual and intercultural education with respect for the cultures and languages of indigenous peoples. The main obstacle to full enjoyment of the right to education has been assimilationist models of education and education systems' ignorance of or failure to appreciate indigenous languages and cultures. In recent years this situation has begun to change, and there are now several countries that officially recognize indigenous cultures and agree on the need for bilingual and intercultural education. Indigenous peoples are demanding recognition of their right to education that is taught in their own language and is adapted to their own culture.

6.2 The Implementation Gap

During the last two decades numerous constitutional and legislative reforms were carried out in many countries through which indigenous peoples and their civil and political rights, and more particularly their economic, social and cultural rights were recognized. Some of these legislative provisions are broader than others; in some cases recognized rights are limited and subordinated to the interests of third parties or wider national interests. In his 2006 report, the Special Rapporteur drew attention to two types of problems in such a situation; firstly, there are many cases in which legislation on indigenous issues is inconsistent with other laws. Secondly, in most documented constitutional reforms there is a delay in the adoption of statutory and secondary laws. The main problem, however, is the 'implementation gap' that is, the vacuum between existing legislation and administrative, legal and political practice. This divide between form and substance constitutes a violation of the human rights of indigenous peoples. To close the gap and narrow the divide is a challenge that must be addressed through an adequate human rights policy and focused programs of action. When the HRC asked me to stay on an additional year because the Council had not yet fully reorganized itself, I made a country mission to Bolivia and prepared the final report of my mandate, this time focusing not as much on ongoing human rights violations as on human rights based development and best practices that several UN resolutions had called upon over the years.

My general thematic reports to the Human Rights Council were presented together, on each occasion, with the reports of the official country visits carried out during the year, which included Guatemala and Philippines (2002), Mexico and Chile (2003), Colombia and Canada (2004), South Africa and New Zealand (2005), Ecuador and Kenya (2006), and Bolivia (2007). I also attended follow-up meetings in Philippines, Guatemala and Canada, organized by local institutions some time after my initial visit, which were intended to evaluate the results and local impact of my earlier missions. My country reports were annexed to the annual thematic report.

The eleven country mission reports, the three evaluation reports and several other, non-official visits to other countries, provide a good overview of the human rights situation of indigenous peoples in the 7 years (2001–2007) of my mandate as special rapporteur. The country reports include specific recommendations to distinct actors (governments, indigenous peoples, international agencies etc.) whereas the recommendations in the thematic reports are of a more general nature. At the time of the presentation of my annual report to the Commission/Council, the delegations of the countries concerned had already received my country reports and had usually prepared written statements about them. Their interventions were usually very cordial and supportive of the SR's work, but they might also point out certain points on which they disagreed. In one case only during all those years did a state representative express outright hostility to me personally and question my good faith, professional standards and moral integrity. He was immediately rebutted by an indigenous organization from his own country which strongly supported my report, and later I received a letter of apology from the country's presidential office. As country reports are submitted to their government for comments prior to the final draft, there were always minor details or corrections to be made in the text. But as the full responsibility of the report belongs to the expert himself, I was surprised to receive, from another country, an almost entirely rewritten report in which probably an official of the foreign ministry or some other department included his/her own ideas regarding indigenous rights. Obviously, I had to disregard this unwarranted interference, which was surely offered in good faith.

The Council session's time table would allow for a few minutes of 'debate' on my annual report which was usually limited to a few questions and answers with a small number of government delegates. As far as I am aware, the Council did not act any further on my proposals and recommendations. In time, I was overcome by a growing sense of frustration at not being able to follow-up on my country visits and reports. Occasionally, I was informed that in one or the other country certain measures had been or were being taken to put into practice one or several of my recommendations. But the general feeling I had (and still have) is that things continue more or less the same regardless of a country visit. Still, certain positive results were achieved and this is because indigenous and other human rights organizations were able to use my reports as one more instrument in their struggle for human rights. In their hands, these reports were sometimes very useful when lobbying or negotiating with the authorities or making their claims widely known by the public, and are frequently quoted in public debates. The UN, as is well known, does not possess any enforcement mechanisms of its resolutions, and this is particularly so in the field of human rights. As the saying goes among the delegates and the specialists, the best we could hope for was 'blame and shame'. But this also has its limits. As time goes by, when human rights violations occur, although the blame remains, states are becoming increasingly immune to the 'shame' that goes with it.

After the adoption of UNDRIP the Council instructed the Special Rapporteur to also promote the Declaration and further its implementation. In 2007 my second term ended and a new special rapporteur, Prof. James Anaya, a well known human rights lawyer, was appointed. His work has contributed precisely in this direction, in making the Declaration better known and helping convert it into a strong and effective international instrument in the cause of the human rights of indigenous peoples. The satisfaction remains that indigenous peoples who have long struggled in vain for their inherent rights, now have in the UN Declaration a most important instrument for their recognition and protection, the full implementation of which still lies in the future. The mandate of the Special Rapporteur is a small but significant contribution to this process.

Rodolfo Stavenhagen being appointed an honorary elder of the Ogiek tribe in Kenya. *Source* Personal photographic collection of the author

Chapter 7
A Report on the Human Rights Situation of Indigenous Peoples in Asia (2007)

During my mandate as Special Rapporteur on the Rights of Indigenous Peoples of the United Nations, I was requested by the UN Permanent Forum on Indigenous Issues to prepare a report on the situation of indigenous peoples in Asia. Based on specialized documentation, direct information provided by indigenous people, academics and governments, and on data given to me during my visits to various countries in the region, I presented this report to the United Nations in 2007. (This text circulated as UN document: E/C.19/2007/CRP.11 (15 May 2007), Permanent Forum on Indigenous Issues Sixth Session, New York, 14–25 May, 2007, Item 6 of the provisional agenda. Half day discussion on Asia. General considerations on the situation of human rights and fundamental freedoms of indigenous peoples in Asia presented by the United Nations Special Rapporteur on the situation of human rights and fundamental freedoms of indigenous people, Mr. Rodolfo Stavenhagen. This text is in the public domain).

Abstract This report presents a general overview of the situation of the rights of indigenous peoples in Asia, based on the information gathered by the Special Rapporteur from various sources during recent activities in the region, including activities organized by the Office of the High Commissioner for Human Rights in Cambodia and Nepal, a follow-up visit to the Philippines, and the First Asian Regional Consultation with the Special Rapporteur. Indigenous peoples in Asian countries face similar patterns of discrimination and human rights violations as in other parts of the world. Drawing from specific examples in various Asian countries, the report focuses on issues of particular concern in the region, including the steady loss of indigenous lands, territories and natural resources; situations of internal conflict, violence and repression faced by these peoples, the implementation of peace accords and autonomy regimes, and the special abuses faced by indigenous women.

7.1 Introduction

The mandate of the Special Rapporteur on the situation of human rights and fundamental freedoms of indigenous people was established by the Commission on Human Rights in resolution 2001/57, extended for a further period of 3 years in 2004 (resolution 2004/62) and renewed by the Human Rights Council in 2006 (decision 1/102). According to his mandate, the Special Rapporteur is expected to

R. Stavenhagen, *Peasants, Culture and Indigenous Peoples*,
Texts and Protocols 4, DOI: 10.1007/978-3-642-34153-3_7, © The Author(s) 2013

"gather, request, receive and exchange information from all relevant sources [...] on violations of [indigenous peoples'] human rights and fundamental freedoms", and to "formulate recommendations and proposals on appropriate measures and activities to prevent" these violations. The present report is submitted in accordance with the decision taken by the Permanent Forum on Indigenous Issues at its fifth session to devote half a day to the discussion during its next session to discuss the issues of indigenous peoples in Asia.

The situation of the human rights of indigenous peoples in Asia raises concerns at different levels. They are discriminated and victimized for their origin and identities, especially in the case of women. They are excluded from full participation in the political life in the countries in which they live. They remain at the margin of national development efforts, and they score low in all indicators in relation to their enjoyment of basic rights such as education and health. They are impoverished as a result of the loss of their traditional lands, territories and lifestyles. They suffer from violence as a result of the defense of their human rights, often by the authorities of their own countries. While these processes are experienced by most indigenous peoples around the world, the situation of indigenous peoples in Asia presents a number of specificities.

These initial considerations on the human rights of indigenous people in Asia, based on the recent activities of the Special Rapporteur, do not attempt to provide a full picture of the situation. These activities include the National Consultation with the Special Rapporteur organized by local indigenous organizations and NGOs, which took place in Quezon City, Philippines, on 2–3 February 2007; the Seminar on Indigenous Peoples and Access to Land in Cambodia, organized by the Office of the High Commissioner for Human Rights (OHCHR), the International Labour Office (ILO), and the United Nations Development Programme (UNDP); and the NGO Forum on Cambodia, an and the NGO Forum on Cambodia, and the First Asian Regional Consultation with the Special Rapporteur, organized by Tebtebba and the Asia Indigenous Peoples Pact Foundation, which took place in Phnom Penh, Cambodia, on 7–8 and 9–11 February 2007, respectively; and the various meetings and on-site visits to communities organized by OHCHR in Nepal, on 23–27 April 2007.

7.2 Indigenous Peoples in Asia

Indigenous peoples in Asia are among the most discriminated against, socially and economically marginalized, and politically subordinated parts of the society in the countries where they live. Time and again disregarded in State's law and policy, they number an estimated 100 million people distributed in virtually all Asian countries, often across State borders. Their traditional territories are frequently found in remote areas where they have historically resisted the drive of colonization and nation-building, including some of the most bio-diversity rich areas of the world. The push of globalization and the State development policies in recent decades have however endangered the continuation of their traditional lifestyles, and they are victims of serious human rights violations as a consequence of the dispossession of their lands and natural resources, widespread violence and repression, and assimilation

Asian States differ in the legal recognition and status granted to indigenous peoples in their own countries, and also in the terminology applied to refer to these different groups in their domestic policies and legislation. Thus, depending on the specific country, they are sometimes referred as 'tribals' or 'tribal people', 'hill tribes', 'scheduled tribes', 'natives', 'ethnic minorities', 'minority nationalities' and other similar denominations. Specific terms are also used in national languages, like Adivasis (original inhabitants) in India and Bangladesh, Orang Asli (original peoples) in Malaysia, or Janajata in Nepal.

In colonial times, some indigenous peoples were given special legal status, like in Bangladesh, India, Indonesia, Malaysia and Myanmar. After independence, however, many Asian countries asserted the principle of 'national unity' to suppress any specific recognition of indigenous peoples as such, but this approach has begun to change in recent years. In a number of countries, indigenous peoples are granted constitutional recognition or are the object of special laws, as in the Constitution of India (1950) (referring to indigenous peoples or adivasis as 'scheduled tribes'); the Constitution of Malaysia (1957) (including special provisions in relation to the 'natives' of Sabah and Sarawak); the Indigenous Peoples' Rights Act (IPRA) of the Philippines (1997); and the Cambodian Land Law (2001). Nepal passed in 2002 the National Foundation for the Development of Indigenous Nationalities Act (NFDIN Act), and indigenous peoples are recognized in the 2006 interim Constitution. The Constitution of Pakistan (1973) recognizes federally and provincially administered Tribal Areas, and involves tribal authorities in decision-making in these areas. In other countries, indigenous peoples are referred to as ethnic minorities and given a legal treatment similar to that of other minority groups, like in the cases of China, Vietnam, or Laos. In other countries, while not explicitly recognized as different collectivities, indigenous peoples may have a distinct legal status. In Indonesia, most peoples who fall under customary law (Adat) self-identify as indigenous peoples. In Japan, the Ainu are not officially considered as indigenous peoples in the 1997 Ainu Cultural Promotion Law, but a number of court decisions have affirmed their rights based on international indigenous rights standards. This is also the case of Malaysia, where the courts have affirmed the aboriginal title of the Orang Asli over their traditional lands.

In addition to the recognition in domestic legislation, three Asian countries, India, Bangladesh and Pakistan, are parties of the 1957 ILO Convention on Indigenous and Tribal Populations in Independent Countries (No. 107), and they report regularly on the implementation of the convention to the ILO Committee of Experts. Nepal has recently started the procedure to ratify the successor instrument, the 1989 ILO Convention on Indigenous and Tribal Peoples in Independent Countries (No. 169), and will thus become the first Asian country to have ratified this important instrument. Moreover, the situation of indigenous peoples in Asian countries is now routinely examined by United Nations treaty bodies in relation to the implementation of the State's general international human rights obligations.

Despite these varied denominations and legal treatment, some States still oppose the relevance of the discussion on the rights of indigenous peoples in the Asian context. Regardless of the controversy around issues of definition, there is an overarching consensus among Asian legal and political actors on the need to address the human rights issues faced by these groups as a result of their distinct identities, lifestyles, and histories. These issues are very similar to those faced by indigenous peoples in other parts of the world, and fall entirely within the sphere of the current international concern on the rights of these peoples, as reflected, inter alia, in the United Nations Declaration on the Rights of Indigenous Peoples. As pointed out by the Committee on the Elimination of Racial Discrimination (CERD), the Governments concerned should provide for the protection of indigenous peoples' rights as recognized by international law, "regardless of the name given to such groups in domestic law" (CERD/C/LAO/CO/15, para. 17). From this perspective, this report will analyze the main trends regarding the situation of the rights of indigenous peoples of Asia, putting a special emphasis on the issues of most immediate concern.

7.3 Issues of Special Concern Regarding the Rights of Indigenous Peoples in Asia

7.3.1 The Loss of Indigenous Peoples' Lands and Territories

Some of the most serious forms of human rights violations that indigenous peoples' experience all over Asia are directly related to the rapid loss of indigenous lands and territories, a process that, while affecting indigenous peoples all over the world, is particularly marked in the Asian context. Development projects, plantation leases, logging concessions, and the establishment of protected areas have been major forces in the increasing loss of indigenous lands, leading to the massive displacement of indigenous peoples form their traditional territories, the degradation of their traditional environment, and rising poverty and migration. This trend is fostered by the absence in many Asian countries of precise legal regulations affirming indigenous peoples' customary rights over their traditional lands, territories and resources, as well as by the lack of adequate consultation procedures in relation to development projects taking place in indigenous territories.

In Thailand, despite the recognition of customary natural resource management by local communities, legal instruments adopted in recent years, such as the Land Act, the National Reserve Forests Act or the National Parks Act, have failed to recognize indigenous and tribal peoples' traditional land tenure and use patterns. The enforcement of these laws have resulted in the expulsion of many indigenous and tribal peoples, considered to be illegal encroachers on their ancestral lands, as well as in a number of unresolved disputes between state lands (including national parks, watershed areas and forestry preservation areas) and community lands. Corruption by law enforcement officers related to the forest industry is said to be rampant.

The development of single-crop, export-oriented plantations has involved the destruction of the natural habitat in both highlands and lowlands where indigenous peoples live, severely limiting the amount of land available for their livelihood and depleting water sources. Only in Sarawak (Malaysia), an estimate of 2.4 million hectares have been given under plantation licenses for the monoculture of palm oil and pulp. Many of these concessions are given over indigenous traditional lands declared 'development areas' and leased for prolonged periods. Indonesia has announced its intention to become the world's largest producer of oil palm, seen as a blooming alternative source of energy, and the official target is to plant 4.6 million hectares throughout the archipelago. This has justifies the transformation of the remaining forest areas into large plantations, with devastating effects on the local indigenous communities.

Land grabbing in Cambodia has became a dramatic example of a trend that is also discernable in other Asian countries. Even though the 2001 Land Law incorporates a number of advanced provisions concerning indigenous communal lands, indigenous communities are losing their lands at an alarming rate as a result of economic concessions, illegal land transfer, and widespread Government corruption. This dynamic is mounting in the densely indigenous-populated provinces of Ratanakiri and Mondulkiri, where the dispossession of indigenous lands has resulted in increased rates of poverty and forced migration. Only in the last decade, an estimated 6.5 million hectares of forest have been expropriated through concessions to timber companies, and another 3.3 million hectares were declared protected areas (see the Special Rapporteur's last thematic report, A/HRC/4/32, para. 15). This critical situation is fostered by the insufficient legal development of the indigenous land provisions of the Land Law, including the lack of a procedural framework for land demarcation and titling; many observers claim that there will be little land left to title by the time the sub-decree on titling is really implemented. The Special Representative of the Secretary-General for Human Rights in Cambodia has repeatedly called attention to the seriousness of the situation, and has recommended that until the adoption of the sub-decree on collective ownership of indigenous lands, a moratorium on land sales affecting indigenous peoples should be considered by relevant authorities (E/CN.4/2006/110, para. 82 (h)).

In the Philippines, the *Indigenous Peoples Rights Act* (1997), recognizes indigenous peoples' rights over their ancestral lands and territories, and incorporates a process of demarcation and titling through the granting of *Certificates of Ancestral Domain Titles* (CADT). In the last 6 years, more than 670 CADT applications have been submitted. With an average of 4.5 titles issued per year, it has been estimated that the *National Commission on Indigenous Peoples* will take almost 25 years to issue titles over the existing applications. Among the reasons of the slowness of the titling process, the existence of overlap between ancestral domain areas and existing leases for mining, agro-forest, logging and pasture have been noted.

The loss of access to natural resources is similarly experienced by coastal peoples. For instance, the Palawan and the Molbog tribes in Bugsuk, Southern Palawan, are still struggling to regain access to their ancestral marine territory after a pearl farm was established. Fishermen who are caught in the perimeter of the farm complain about harassment, ill treatment and illegal detention by company guards. Confronted with these vested interests, the National Commission has been accused of a weak commitment towards fully implementing its mandate. In the report on his visit to Japan, the Special Rapporteur on contemporary forms of racism described how the Ainu are still greatly limited in their capacity to fish salmon, their traditional food. This situation is "humiliating, since it puts them in a position of dependence on the public authorities in the access to their ancestral alimentary resources" (E/CN.4/2006/16/Add.2, paras. 45–47).

7.3.2 The Situation of Forest Peoples

Commercial logging, both illegal and Government-sponsored, is a major source of indigenous land loss in practically all countries of the region. For instance, in Bangladesh, India, Indonesia, the Philippines and Thailand, forests are considered State-owned lands, and indigenous communities lack any legal venue to counter Government policies in these areas or seek compensation in cases in which their traditional lands are lost.

The *Andhra Pradesh Community Forest Management Project* (APCFMP), launched in 2002 in India with the support of the World Bank, has been opposed by Adivasi organizations, who claim that the procedural safeguards incorporated by the World Bank (including the establishment of forest protection committees or Vana Samrakshana Samithi) have not been adequately implemented.

In Malaysia, indigenous communities have denounced that the national forestry certification system run by the *Malaysian Timber Certification Council* (MTCC) fails to recognize and protect indigenous customary rights over the forest they have traditionally occupied or used for their subsistence. Several cases have been brought to the national courts as a result of the granting of timber certification to private companies operating in communal lands, without prior consultation of the communities concerned and with no compensation paid to the people. In some cases, indigenous communities have mobilized against logging in their ancestral territories, like the Dusun community of Terian, Sabah, which recently stopped an illegal logging road that threatened its traditional forest near Crocker Range National Park. Similarly, the Penan people in the Middle Baram region of Sarawak, who have led several peaceful blockades and have endured violence by loggers and security forces.

As in other parts of the world, indigenous peoples in Asia have suffered the direct consequences of the establishment of national parks. This is for instance the case of the Modhupur National Park Development, in Modhupuar, Tangail District (Bangladesh). The Eco-Park project, initiated in 1999, involved the erection of walls that cut across the Modhupur forest, ancestral land of the Garo and Koch peoples, without previously consulting them. Suspended in 2004, the Eco-Park project was resumed after the declaration of the state of emergency in January 2007, and there have been serious allegations of the detention of indigenous leaders, torture and even killings.

Despite international praise for its international conservation efforts, Nepal's community forests have forced many indigenous communities, like the Chepangs and the Rautes, from their traditional lands. In Sri Lanka, the Wanniyala-Aetto indigenous people were evicted in 1983 from the lands which they have occupied for centuries to give way to the Maduru Ova National Park; since then, their number has fallen to only 2,500 members, half of the original population, and they are on the verge of virtual extinction. More than 1,000 Adivasis have been expelled from the Muthanga Wildlife Sanctuary in Wayanad, State of Kerala, India. In Indonesia, the Moronene people of Southeast Sulawesi have been evicted several times since their traditional territory was declared a conservation forest in 1997. A similar case is that of the Wana people after the government announced the creation of the Morowali conservation area in their traditional territory. The Semi tribe, in Malaysia, is opposing the establishment of a National Botanical Garden in the Perak State, a project that aims at becoming a major tourist attraction but that would expel the community from the ancient rainforest in which they lived for generations, and over which they do not possess a formal title.

In recent years, a number of countries have started to address the legal vacuum concerning indigenous peoples' communal land rights with the adoption of new legislation. Following the example of countries like Cambodia or the Philippines, the 2003 Land Law in Vietnam includes the category of 'communal land', which has opened the possibility for indigenous people to apply for titles over their ancestral land and forest rights; some difficulties still need to be clarified concerning the interpretation of various provisions of the law. In 2006, after many massive protests by Adivasis and forest dwellers, India adopted the Scheduled Tribes and other Traditional Forest Dwellers (Recognition of Forest Rights) Bill. The bill grants extensive rights to indigenous forest dwellers, including the right to possess forest land for habitation and self-cultivation purposes, as well as the right of access to forest resources and to participate in conservation efforts. The Bill further incorporates a special procedure for the establishment of 'critical wildlife areas', as well as for the informed relocation and rehabilitation of the affected communities.

In the absence of specific legislation, national courts have played a major role in affirming indigenous peoples' rights over their traditional forest. For instance, in Malaysia, a number of decisions by the Supreme Court, including the path-breaking Sagong Tasi v. Negeri Kerajaan Selangor (2002), have recognize the existence of Orang Asli's native title over their traditional lands even in the absence of a formal title deed, despite the lack of statutory recognition of their rights in Malaysian law.

7.3.3 Forced Relocation and International Resettlement

One of the most serious threats to indigenous peoples' survival in Asia relates to the construction of megaprojects and other forms of forced relocation or resettlement in the name of 'national development', which take place in several Asian countries at a particularly alarming rate. The Special Rapporteur has expressed his concern in relation to some of these projects.

In India, according to the 5-Year Plan (2002–2007) of the National Commission on Scheduled Castes and Scheduled Tribes, 8.54 million tribals have been displaced from their traditional lands as a result of development projects in the states of Andhra Pradesh, Bihar, Gujarat, Maharastra, Madhya Pradesh, Rajasthan and Orissa, of which less than a quarter have been resettled. According to the Commission, this massive displacement has led to "loss of assets, unemployment, debt bondage and destitution". The Special Rapporteur, as well as other human rights mechanisms have repeatedly expressed their major concern about the Sardar Sarovar Dam and Power Project, a multiyear, mutipurpose project affecting areas in the states of Gujarat, Rajasthan, Madhya Pradesh and Maharashtra, involving the relocation of 320,000 people and affecting the livelihood of thousands of others. There is concern about the lack of adequate compensation or resettlement schemes of the tribal communities affected. In addition, 168 new dams are scheduled for construction in north-eastern India, without the meaningful participation by and the consent of the Bodos, Hmars, Nagas and other indigenous communities that have traditionally owned the land. These dams, it is argued, that will provide electric power to other parts of India, will create irreparable harm to indigenous peoples' traditional subsistence communities. Concern has also been expressed that these proposed dams are located in a highly seismic area.

Similar large scale displacement has resulted from mining. The Government of Jharkhand has open lands to 41 steel and mining companies for large scale resource extraction, which will result in the destruction of 57,000 ha of forest and in the displacement of 9,615 families, 80 % of whom belong to scheduled tribes. Similarly, State-sponsored mining projects in Orissa have resulted since 2004 in the displacement of hundreds of Jarene families, and 300 other families are still under threat as a result of new projects. The Khasi people of Eastern Meghalaya now face the proposed resumption of uranium mining in its traditional territory, involving the displacement of an estimate of 30,000 people, the massive influx of non-indigenous settlers, and possible health risks.

The 13 dam cascade project on the Chinese portion of the Nu river would have a considerable effect on the Nu, Lissu, Yi, Pumi and other ethnic minorities in the area, and its impact of the biodiversity-rich Three Parallel Rivers World Heritage Site has raised the concern of UNESCO's World Heritage Committee. Vietnam is currently embarked in the construction of the Son Lam Dam, the largest such project in the region, involving the submersion of 24,000 ha of land and the forced removal of 100,000 people, mostly ethnic minorities. The Bakun Dam in Malaysia is reported to cause the forced displacement of 5,000–8,000 indigenous persons from 15 communities by clear-cutting 80,000 ha of rainforest. In Laos, the construction of the Nam Theun 2 dam, in Khammouane province, involves the displacement of as many as 6,200 indigenous people. The Special Rapporteur, along with other special procedures, is currently engaged in a constructive dialogue with the Government of the Lao People's Democratic Republic, the World Bank and other donors, promoting the effective implementation of the relocation and compensation program.

Laos and Thailand have undertaken the resettlement of many tribal people as part of their program of eradication of drug plantations. The Government of Thailand launched in 2003 a *Master Plan for Community Development, Environment, and Narcotic Plant Control on the Highland*, leading to the displacement of indigenous communities. Due to the relocation schemes, many of these communities have broken up, and they often lack alternative ways to provide for their subsistence. The Lao Government's campaign of eradication of opium poppy has been internationally praised as a success, but it has led the displacement of an estimated 65,000 hill tribe people into new villages where they are said to experience severe food shortages, disease, and mortality rates as high as 4 %.

The Vietnamese Government has adopted a 'Fixed Field/Fixed Residence' policy that involves the resettlement of ethnic minorities, including many indigenous and tribal communities, from remote areas into other more easily accessible locations. The purpose of this resettlement is to make social services more easily available to these communities, but also to replace their traditional slash-and-burn agriculture, viewed as inefficient by the Government, by other methods of sedentary agriculture. The resettlement has generated the social and cultural disintegration of many of these communities, as well as increased ethnic tension as a result of a state-sponsored migration program to bring non-indigenous settlers into the indigenous highlands. A similar stand has been taken in Laos, where numerous Hmong communities have been forcibly relocated by the Government from their traditional lands in the highlands and resettled in so-called 'focal sites', together with other ethnic minority group or Hmong from different clans. Reports indicate that these resettlement sites are often not arable lands, and that their traditional life has been eroded. In its last concluding observations on Laos, CERD recommended the Government to avoid displacement, and if necessary, to "ensure that the persons concerned are made fully aware of the reasons for and modalities of their displacement and of the measures taken for compensation and resettlement" (CERD/C/LAO/CO/15, para. 18).

The Dukha (Tsaatan) people, a reindeer-herder community living in Mongolia's Darhat Valley, endured similar attempts of forceful relocation during the 1950s. Now they are striving to retain their traditional culture against the depletion of their herds and the loss of their traditional lands. The establishment of the Lake Baikal and Sayan Mountains Peace Park, in the border between Russia and Mongolia, home of the Dukha and other peoples like the Soyot and Buryat, or the 2002 adoption of the Charter Agreement on the Protection of the Transboundary Reindeer Herding Cultures of Russia and Mongolia, constitute important initiatives to promote the respect for indigenous peoples' semi-nomadic lifestyles with the protection of the environment in their traditional territories.

7.3.4 Conflict and Repression

Historically, the denial of equal enjoyment of political and other rights has led to an increase in violence that, in many cases, has involved indigenous peoples directly. Internal conflict has posed an enormous burden on indigenous communities and other parties involved, and have sometimes led to massive human rights violations. Countless cases are also reported concerning abuses suffered by indigenous peoples by military and paramilitary forces in the name of public security, anti-insurgency, and counter-terrorism. Examples of these dynamics in the past decades include the armed insurgencies in north-eastern in India, in Aceh and West Papua, in Indonesia, and in Mindanao, in the Philippines, as well as the protracted conflicts in Myanmar and Nepal. In Laos and Vietnam, some indigenous peoples still face retaliation for their involvement in armed conflicts during the American War a generation ago, and they are reportedly denied full citizen rights and persecuted as criminals.

Indigenous peoples (or 'ethnic minorities') in Myanmar, like the Kachin, Karen, Karenni, Mon, or Shan, represent one third of the country's total population. They have endured the worst consequences of the civil war that has stricken the country for half a century, and which involved indigenous groups fighting against the military government. They experience all sorts of human rights violations in the context of counter-insurgency operations against indigenous groups, including extrajudicial killings, massacres, torture and sexual violence, and large movements' refugees and internally displaced persons as a result. The ILO has also denounced the practice of forced labor, particularly in indigenous areas.

Different sources have documented the countless deaths of civilians, including children and elders, as a result of the continuous struggle of the Hmong with the Lao Government since 1975. It has been estimated that 20 rebel groups are surrounded by Lao military and reduced to starvation and disease in the forest where they have sough refugee. Many of them have fled to Cambodia and Thailand, where there have been reports of hundreds of deportations. Following the upsurge of military activity reported in recent years, several hundred Hmong have reportedly 'surrendered' to Lao authorities, and episodes of human rights abuses have been reported, like the killing and gang rape of five girls by armed forces in 2004 (CERD/C/LAO/CO/15, para. 22).

The Special Rapporteur has received reports documenting hundreds of human rights violations of individual Degar (Montagnard) people in Vietnam. These allegations refer to cases of arbitrary arrest, ill treatment, torture and extrajudicial killing by security forces. In addition, it has been alleged that 350 Degar prisoners remain in Vietnamese prisons for human rights activism, for spreading Christianity or for attempting to flee to neighboring countries. Following the February 2001 and April 2004 protests in the Central Highland Region of Vietnam, when numerous killings and other human rights abuses by security forces were reported, many hundreds of indigenous asylum seekers fled the country into neighboring Cambodia in fear of Government repression.

The massive scale of political killing of indigenous leaders and human rights defenders in the Philippines has been object of increased international concern in recent years. Leaders and member of indigenous organizations are tagged as 'legal fronts' of the Communists because of their human rights related activities, and also because of their opposition to mining operations and other megaprojects that threaten indigenous communities. The Melo Commission, established in 2006 by the Parliament to investigate the situation, concluded that the majority of the killings could be attributed to members of the Philippine military. According to a report of Indigenous Peoples Watch-Philippines, 119 such killings took place in the period April 2001–January 2007. Recent examples of such acts are the killing of Rafael Markus Nagit, in June 2006, and the attempted assassination of Dr. Constancio 'Chandu' Claver on July 2006, leading to his wife's death. The situation has been reported on by the Special Rapporteur during his official visit to the country in 2002 (see E/CN.4/2003/90/Add.3, para. 46); since his visit the murder of another 84 indigenous leaders has been reported.

Indigenous peoples of north-eastern India have repeatedly denounced the human rights violations committed by security forces under the *Armed Forces (Special Powers) Act* (AFSPA) (1958), adopted in the context of an armed conflict in Assam, Nagaland, and Manipur. After a 1997 decision of the Indian Supreme Court that questioned the constitutionality of several of the AFSPA provisions, a review committee appointed by the Government in 2004 proposed the amendment of the Act, but its recommendations were never publicly released, and violations of human rights continue unabated. Following the declaration of the state of emergency by the President of Bangladesh in January 2007, the Special Rapporteur have received many allegations of suppressive actions against indigenous leaders and organizations that would have involved the Joint Forces, consisting of the military, the *Rapid Action Battalion* (RAB), the *Bangladesh Rifles* (BDR), the police and intelligence servicemen, which were given special powers to control corruption. Among the alleged abuses, there are reported cases of arbitrary arrest, detention and torture of members of Jumma leaders in the Chittagong Hill Tracts and other regions. Decades of conflict in Nepal and in several Indonesian provinces, including Aceh (Nanggröe Aceh Darussalam) and West Papua (Irian Jaya), have left behind a tragic record of killings, forced displacement and other serious human rights abuses among local indigenous groups. Indigenous peoples now demand full participation in the post-conflict political arrangements, and plead for transitional justice schemes to repair past human rights violations.

Local conflicts resulting from the lack of recognition of the rights of indigenous peoples to their communal lands is another permanent source of repression and abuse and often leads to violations of human rights violations of indigenous peoples. The Special Rapporteur has received many reports from countries such as India, Indonesia, Laos, Malaysia and Thailand, of arbitrary arrest or fake criminal charges made against members of indigenous and tribal peoples, as well as other forms of threats and intimidations, as a result of their mobilization to defend their rights against State authorities. Cases of ill-treatment and torture during detention, as well as extrajudicial killings have also been widely reported. In India, for instance, 15 Adivasis were killed in 2003 as a result of the use of excessive police force in the demonstrations to protest against the establishment of the Muthanga Wildlife Sanctuary. In Laos, 10 Degar people were killed as a result of the 2004 protests in the Central Highlands. In the Philippines, the lethal conjunction of militarization and large scale mining and dam projects have led indigenous peoples to coin the expression 'development aggression', which is to blame for a wide range of human rights violations, including murders, massacres, and illegal detention. The critical situation faced by the various Lumad in Mindanao or the Tumandok on Panay Island, are cases in point.

7.3.5 Citizenship Rights, Refugees and Asylum Seekers

The lack of citizen rights has been a long-standing cause of human rights violations against members of the hill tribes in Thailand since the enactment of the Citizenship/Nationality Act in 1965. According to 2004 estimates, 90,700 original hill people are not given Thai citizenship or any enjoy other legal status, remaining stateless in their own countries. The lack of access to citizenship rights make them subject to many abuses, like charges of illegal entrance in the country and denial of freedom of movement, threats, intimidation, and bribery. They are also denied access to basic social services, including health care and education as well as income generating activities. A mix of discriminatory laws and procedures, deeply-rooted prejudices, and corruption are among the main causes of this situation, which has been repeatedly denounced by human rights bodies, including the *Committee of the Rights of the Child* (CRC/C/THA/CO/2, para. 24), the *Committee for the Elimination of Discrimination Against Women* (CEDAW/C/THA/CO/5, para. 78), and the Human Rights Committee (CCPR/CO/84/THA, paras. 22–24).

Indigenous and tribal peoples in Myanmar face the worst consequences of the civil war that has stricken the country for decades. For instance, as a result of the large-scale offensive that took place in Karen state during 2006, 27,000 civilians were displaced, and some 232 villages destroyed. According to one independent source, between 2004 and 2006, some 470,000 Mon, Karen, Shan and Karenni were internally displaced as a consequence of violence, military operations and human rights abuses. Others have been able to flee the country, and survival in extremely difficult conditions in formal or informal refugee camps in neighboring countries.

Special mention must be made of the plight of the Khmer Krom people in southern Vietnam who complain about serious human rights violations, especially concerning citizenship, religious freedom, land rights and gender issues, as a result of complex historical and geopolitical factors.

7.3.6 Autonomy Rights and Implementation of Peace Accords

In a number of Asian countries, constructive arrangements, including autonomy regimes, have sought to accommodate the ethnic diversity of some regions, or to put an end to decades of armed conflict. Inasmuch many of these arrangements provide for limited autonomy in local affairs, political participation, and land and cultural protection, they represent positive steps towards the promotion of the rights of indigenous peoples. However, comparative experience suggests that these arrangements have a mixed record in terms of implementation, and that much remains to be done by the Governments concerned, and by the international actors committed to the monitoring of these arrangements, to ensure that indigenous communities are actively involved and their human rights concerns taken into account.

Similar dynamics are found in the *Chittagong Hill Tracts* (CHT), in Bangladesh, where an autonomy regime was instituted in 1997 following the Peace Accord between the Government and the Parbatya Chattagram Jana Samhati Samiti, a party representing 11 different indigenous communities of the Jumma people. Indigenous people claim that many vital provisions of the Accord have not yet been put in place, including the setting up of a functioning Land Commission (constituted in 1999 but still not fully operative); the rehabilitation of Jumma refugees and internally displaced persons, and the formation of a CHT-based police force. The policy of Government-sponsored transmigration has dramatically changed the ethnic composition of the region, and Bengalese settlers represent now more than 60 % of the region's population, compared to only 2 % in 1947. This influx has facilitated cultural assimilation, while creating increased ethnic animosity over diminishing land and resources. Instead of demilitarizing the area, it has been claimed that the Government has continued sending armed forces to the region under the umbrella of the Uttoran (upliftment) and Shantakaran (pacification) programs, allowing for the military intervention in civilian administration and in the establishment of settler villages.

In 2001, Indonesia adopted the Special Autonomy Law No. 21, aiming at finding a solution to West Papua's political status and to bring peace to the province. Similarly, a Memorandum of Understanding between the Government of the Republic of Indonesia and the Free Aceh Movement was signed in 2005, providing for a limited autonomy to Aceh within basic sectors of public affairs, as well as for the right to consultation concerning international agreements for special interest to Aceh. While constructive arrangements have been seen as positive steps, the experience of West Papua after more than 5 years of the entry into force of the autonomy regime is disquieting. The Government has continued promoting the massive arrival of settlers on the island, the region is still heavily militarized, and episodes of repression and abuse in Puncak Jaya and other parts of the highlands have recently been reported.

Since a cease-fire was reached in Nagaland in 1997, the Government of India and several Naga insurgent groups are involved in a peace process seeking to find the political accommodation of the Naga people under the Indian Constitution. The peace process, which follows decades of violent insurgency in various northeastern states, is subject to ongoing tensions due to the resumption of violence, internal rivalries among the Nagas and the animosity of neighboring communities and state governments at the attempt to extend the ceasefire agreement to areas beyond Nagaland. Despite the many difficulties, the Nagas favor a peaceful settlement of the conflict and demand full implementation of the 1997 agreement as a precondition to achieve this goal.

A deeply entrenched system of ethnic and caste-based hierarchy, along with decades of internal conflict, has led to a disproportionate part of indigenous peoples among Nepal's poor. Nepal's indigenous peoples, who represent 37 % of the national population, have denounced that the recently endorsed Interim Constitution fails to provide them with an equal representation in the Constituent Assembly, and they are now demanding a federal republic based on ethnic and regional autonomy.

7.4 The Rights of Indigenous Women and Girls

Gender-based violence has been recurrently used in the armed conflict in Myanmar, where numerous cases of gang-rape, sexual enslavement and killing of tribal women by members of the military have been reported. Although some of these cases have been well documented, the Military has routinely failed to investigate these abuses. In the CHT in Bangladesh, many cases of rape of Jumma girls and women by settlers backed by the military have been denounced, but in many cases the investigation of these cases is hampered by inaction on the part of the military and even of health professionals. In the Philippines, the militarization of many indigenous areas has also resulted in the sexual abuse of women of local indigenous communities. In India, the AFSFA has justified impunity of sexual violence by members of the military against tribal women, sometimes with the argument that they support insurgent groups.

The increasing numbers of indigenous women who have become victims of sexual trafficking and prostitution is of special concern. While systematic data is still lacking, in countries such as Mongolia, Thailand, Myanmar, Nepal, Laos, Cambodia and Vietnam, indigenous women and girls are prime targets for trafficking and exploitation as beggars, sex workers, domestic workers, and even child soldiers. In areas such as Chiang Mai, in Thailand, where there are thousands of indigenous women working as sex workers, 70–80 % of these women are reportedly HIV positive. In other cases, like in Nepal, Indonesia, Bangladesh or the Philippines, indigenous women and girls are forced to leave their communities and search for jobs in other countries.

7.5 Conclusions

In recent years the plight of indigenous peoples in Asia has started to become a specific issue of concern in the international human rights agenda, as well as in domestic legislation and policies. Indigenous issues are increasingly the object of specific attention by several Asian States in key areas such as land rights, cultural protection, autonomy and self-government and development policies, thus signaling an important change of mentality regarding the recognition of cultural difference and its human rights implications. However, there is still an important implementation gap with regard to existing constitutional and legal provisions, and much remains to be done in order to mainstream indigenous rights in policies and the institutional machinery at the national level. These developments are overshadowed by the human rights violations still suffered by indigenous peoples in some countries of the region as a result of internal conflicts and insensitive official policies.

Indigenous peoples in Asian countries face patterns of discrimination and human right abuses similar to indigenous peoples in other parts of the world. Some of the most serious violations are related to the lack of effective protection in domestic laws and policies regarding indigenous rights over their traditional territories, lands and natural resources, as well as to the their right to participate in decisions affecting these lands and resources. This has lead to widespread violations in practically all countries of the region as result of land-grabbing and corruption, forced displacement associated with the extension of plantation economies, the construction of megaprojects, and particularly dam construction and mining; and other State development policies.

Forest peoples are particularly affected by these dynamics of dispossession and removal, as the forests are quickly disappearing as a result of Government-promoted and illegal logging, and other State policies often with disastrous environmental effects. Pastoralist communities similarly confront the loss of their distinct livelihoods and cultures, essential to nomadic herding, which is frequently deemed ‘backward’ and ‘unecological’ in official discourse and policy.

While militarization and State repression are frequently the source of indigenous peoples’ human rights violations in many parts of the world, the recurrent and widespread character of these abuses in Asian countries gives rise to special concern. Decades-long civil conflicts, insurgency movements, political crimes, and other abuses committed in the name of the struggle against terrorism or secessionism have taken a deadly toll in indigenous and tribal communities. Massacres, killings of social activists and human rights defenders, torture, sexual violence, and displacement are still daily realities for many such communities. While the Special Rapporteur acknowledges the complexity of the various contexts in which these violations occur, the seriousness of these violations leads to the conclusion that the indigenous peoples are widely regarded in many countries as ‘backward’, second-class citizens.

A number of constructive arrangements have been put in place in order to accommodate ethnic diversity or to find a peaceful solution to decades-long conflicts. While these initiatives provide important examples of ways in which the principles of State integrity and autonomy can be combined in the Asian context, a common denominator of ongoing experiences is the lack of implementation of existing legal and political arrangements. Militarization, induced migration, unequal development policies, and resulting human rights abuses are questioning the spirit of such arrangements, while fueling the conflicts they seek to prevent.

As elsewhere in the world, the indigenous women of Asia experience accumulated layers of discrimination and marginalization. They are subject to human rights violations as a result of longstanding conflicts and the impoverishment of their communities. Sexual violence, trafficking and labor exploitation are daily realities for many Asian indigenous women in Asia, a problem that is just beginning to be fully understood.

7.6 Recommendations

The protection of the rights of indigenous peoples is a human rights imperative that cannot be subordinated, nor is it contradictory, to the objectives of national unity or development. The Special Rapporteur calls upon Asian States to give priority attention to indigenous issues, regardless of the constitutional and legal status afforded to these groups in their domestic systems, taking into consideration international norms as well as the positive examples found in comparative legislation in Asia and other parts of the world.

Asian States should continue their efforts to enter into dialogue with indigenous peoples in order to work out constructive legal and political arrangements, within a spirit of mutual respect, autonomy, and self-determination. These demands should not be repressed or criminalized, and their basic human rights should be fully respected at all times, including in situations of conflict.

National legislation in Asian countries should incorporate indigenous peoples' property and use rights over communal lands, forest areas, pastures, and other natural resources, with due regard to indigenous customary laws, traditional lifestyles, and cultural values. Where such legislation exists, renewed efforts should be made in order to make indigenous rights effective, and special emphasis should be put on the demarcation and titling of indigenous lands. The systematic removal of indigenous peoples from their traditional lands as a public policy should be halted, and such removal of indigenous peoples from their traditional lands should be regarded as a last alternative and in cases of utmost necessity, and under condition that they be fully compensated.

Indigenous peoples should be involved in decision-making at all levels in the countries in which they live. They should participate in the design and implementation of all policies that may affect them directly, particularly with regard to development projects taking place in their lands and territories.

Asian countries should be actively and constructively involved in international discussions concerning the rights of indigenous peoples, particularly regarding the United Nations Declaration on the Rights of Indigenous Peoples and the future role of the Human Rights Council in the promotion and protection of indigenous rights. Asian states should consider the prompt ratification of ILO Convention No. 169 on Indigenous and Tribal Peoples, particularly those that are already party of the previous ILO Convention No. 107.

International organizations and agencies, as well as international financial institutions, should mainstream indigenous rights into their programs and activities in Asian countries, on the basis of international norms and their own policy guidelines in this area, irrespective of the level of recognition of these rights in domestic legislation and policies. OHCHR country and regional offices in Asia should further strengthen their programs of work the rights of indigenous peoples, particularly of indigenous women. UNDP and ILO should continue their efforts to promote their policies on indigenous peoples. The World Bank, the Asian Development Bank, and bilateral donors should ensure that their safeguards and guidelines in relation to indigenous peoples are fully respected in their Asian projects.

Rodolfo Stavenhagen visiting an indigenous community at a proposed dam site in La Parota, Guerrero, Mexico, 2008. *Source* Personal photographic collection of the author

Chapter 8
Report on the Impact of Megaprojects on the Rights of Indigenous Peoples (2003)

Abstract To provide an additional view of some of the issues that were dealt with in my reports to the United Nations as Special Rapporteur on the Rights of Indigenous Peoples, I include in this volume extracts of my second and seventh (last) annual reports to the Human Rights Council, dealing, respectively, with the impact on indigenous human rights of large development projects, such as dams, and with so-called 'best practices' carried out governments and other agencies under different circumstances, to implement some of the recommendations that I made during my 7 years as Special Rapporteur.

8.1 Introduction

On 24 April 2001, at its fifty-seventh session, the Commission on Human Rights adopted resolution 2001/57 in which it decided to appoint, for a period of 3 years, a special rapporteur on the situation of human rights and fundamental freedoms of indigenous people with the following functions: (a) to gather, request, receive and exchange information and communications from all relevant sources, including Governments, indigenous peoples themselves and their communities and organizations, on violations of their human rights and fundamental freedoms; (b) to formulate recommendations and proposals on appropriate measures and activities to prevent and remedy violations of the human rights and fundamental freedoms of indigenous people; (c) to work in close relation with other special rapporteurs, special representatives, working groups and independent experts of the Commission on Human Rights and of the Sub-Commission on the Promotion and Protection of Human Rights.

This text was circulated by the UN as document E/CN.4/2003/90 and is in the public domain.

Texts and Protocols 4, DOI: 10.1007/978-3-642-34153-3_8,

On April 15, 2002, the Special Rapporteur, Mr. Rodolfo Stavenhagen presented his first annual report to the Commission (E/CN.4/2002/97), in which he indicated some of his future activities. He is now pleased to present this second annual report to the Commission on Human Rights.

During the time elapsed since the termination of his first report, the Special Rapporteur has continued gathering information on the situation of the human rights of indigenous peoples, following developments in the United Nations system, participating in international and national level conferences and research seminars, evaluations, training workshops and the like, that deal directly with the issues of his mandate, and he has undertaken research on some of the major issues affecting indigenous peoples which he laid out in his first report. (Ibid. para. 113). He has also carried out two official country missions to Guatemala (September 2–12, 2002) and the Philippines (December 2–11, 2002). The country mission reports are available separately (E/CN.4/2003/90/Add.2 and E/CN.4/2003/90/ Add.3). Moreover, he has visited some additional countries to observe the situation of indigenous peoples, in connection with other activities, including Botswana (January 2002), Mexico (April 2002), and Japan (November 2002).

8.2 The Impact of Large-Scale or Major Development Projects on the Human Rights and Fundamental Freedoms of Indigenous Communities

By 'major development project' should be understood a process of investment of public and/or private, national or international capital for the purpose of building or improving the physical infrastructure of a specified region, the transformation over the long run of productive activities involving changes in the use and property rights of land, the large-scale exploitation of natural resources including sub-soil resources, the building of urban centers, manufacturing and/or mining, power, extraction and refining plants, tourist developments, port facilities, military bases and similar undertakings. The purpose of such projects may vary, from furthering economic growth to flood control, generating electric and other energy resources, improving transportation networks, promoting exports to obtain foreign exchange, create new settlements, ensure national security, and generate employment and income opportunities for the local population.

Indigenous peoples live mainly in rural environments. Wherever they have been able to maintain their community lifestyles and their traditional cultures, it is because the areas in which they live have been spared major upheavals resulting from rapid economic and ecological transformations. But this situation has changed rapidly over the last few decades, as national governments, large corporations and multilateral financing agencies turn their attention to so-called undeveloped regions in order to extract natural resources, establish plantations and industrial plants, develop tourist activities, ports, communication hubs or urban centers, build transportation networks, multipurpose dams, military bases or toxic waste dumps. Wherever such developments occur in areas occupied by indigenous peoples it is likely that their communities will undergo profound social and economic changes that are frequently not well understood, much less foreseen, by the authorities in charge of promoting them. Large-scale development projects will inevitably affect the conditions of living of indigenous peoples. Sometimes the impact will be beneficial, very often it is devastating, but it is never negligible.

Traditionally few governments have taken the rights and interests of indigenous peoples into account when making plans for major development projects. As the projects mature, which may take several years depending on their characteristics, the concerns of indigenous peoples, who are seldom consulted on the matter, take a back seat to an overriding 'national interest', or to market-driven business objectives aiming at developing new economic activities, maximizing productivity and profits. For a long time, multilateral financing agencies involved in the planning and execution of such projects appeared to go along with this approach. Hence, the social and environmental concerns expressed by many people, including indigenous communities, have not been given the necessary attention.

In recent years, this situation is changing, as multilateral agencies, national governments and the business community takes up a new interest in indigenous concerns. At the international level, ILO's Indigenous and Tribal Peoples Convention 169 stipulates that:

> The peoples concerned shall have the right to decide their own priorities for the process of development as it affects their lives, beliefs, institutions and spiritual well-being and the lands they occupy or otherwise use, and to exercise control, to the extent possible, over their own economic, social and cultural development. In addition, they shall participate in the formulation, implementation and evaluation of plans and programs for national and regional development which may affect them directly…

Governments shall ensure that, whenever appropriate, studies are carried out, in co-operation with the peoples concerned, to assess the social, spiritual, cultural and environmental impact on them of planned development activities. The results of these studies shall be considered as fundamental criteria for the implementation of these activities.

Governments shall take measures, in co-operation with the peoples concerned, to protect and preserve the environment of the territories they inhabit" (Art. 7). Numerous international conferences have reaffirmed such rights in one formulation or the other, notably the Rio Earth Summit (1992) and the Johannesburg World Summit on Sustainable Development (2002). The World Bank is in the process of adopting a new operational policy that establishes the need to involve indigenous peoples in development projects that may affect them, and the Inter-American Development Bank has laid down similar guidelines for its own activities. Several states have likewise adopted legislation in the same sense.

None have been more concerned with these important issues than indigenous peoples themselves. One recent study reports on "the disproportionate impacts that indigenous peoples suffer from development programs, so long as their human rights are not fully recognized, and so long as they continue to be marginalized in decision-making affecting their lives". Further, indigenous peoples argue that "as the pressures on the Earth's resources intensify, indigenous peoples bear disproportionate costs of resource-intensive and resource-extractive industries and activities such as mining, oil and gas development, large dams and other infrastructure projects, logging and plantations, bio-prospecting, industrial fishing and farming, and also eco-tourism and imposed conservation projects". On the specific issue of large dam construction (on which this report will concentrate), the World Commission on Dams finds that:

> Large dams have had serious impacts on the lives, livelihoods, cultures and spiritual existence of indigenous and tribal peoples. Due to neglect and lack of capacity to secure justice because of structural inequities, cultural dissonance, discrimination and economic and political marginalization, indigenous and tribal peoples have suffered disproportionately from the negative impacts of large dams, while often being excluded from sharing in the benefits.

To the extent that many of these projects are located on the ancestral territories of indigenous peoples, it is not surprising that they should raise the issue of the rights to land, the right to prior consent about use of this land, the right to participation in the decision-making process regarding the implementation of such projects, the right to share in the potential benefits, and beyond this, the right of indigenous peoples to self-determination. Thus, at the twentieth session of the *Working Group on Indigenous Populations* (WGIP)

> …virtually every indigenous participant stated that their right to self-determination is a pre-condition for the realization of all other human rights, and must be considered as the bedrock that ensures their self-governance, whereby they can participate in decision-making processes in policies that directly affect them. They therefore reiterated the intrinsic link of the right to self-determination to various other indigenous human rights issues such as the right to land and natural resources, the preservation of cultural identity, and the rights to language and education.

The right to free, informed and prior consent by indigenous peoples continues to be of crucial concern, inasmuch as too many major decisions concerning large-scale development projects in indigenous territories do not comply with this stipulation, clearly set out in para. 6 of ILO's convention 169, which provides that governments shall

> (a) consult the peoples concerned, through appropriate procedures and in particular through their representative institutions, whenever consideration is being given to legislative or administrative measures which may affect them directly;
> (b) establish means by which these peoples can freely participate, to at least the same extent as other sectors of the population, at all levels of decision-making in elective institutions and administrative and other bodies responsible for policies and programs which concern them…

Likewise, Article 30 of the UN Draft Declaration on the rights of indigenous peoples also provides that States shall obtain free and informed consent prior to the approval of any project affecting their lands, territories and other resources, particularly in connection with the development, utilization or exploitation of mineral, water or other resources. The Proposed American Declaration on the Rights of Indigenous Peoples (Article 21[2]) contains a similar provision. The importance of the principle of free, prior and informed consent was also highlighted in the recommendation of the UN Workshop on *Indigenous Peoples, Private Sector Natural Resource, Energy and Mining Companies and Human Rights* (5–7 December 2001).

In some states legislation has progressed in this direction. The Aboriginal Land Rights (Northern Territory) Act 1976 (amended in 1987) of Australia not only recognizes the right of Aborigines to own the land, but also provides in effect the right to veto over mining for a 5-year period. Furthermore, a land council with the mandate to represent the interests of Aboriginal land owners may not consent to the grant of a mining interest or construction of a road unless the traditional owners of the land understand the nature and purpose of the proposed mining or road construction proposals as a group and consent to them.

The Indigenous Peoples Rights Act 1997 of the Philippines recognizes the indigenous right to ancestral domain and the land title to traditional lands. Philippine law also requires a developer or company to obtain free, prior and informed consent of indigenous peoples for certain activities, such as (a) exploration, development and use of natural resources; (b) research-bioprospecting; (c) displacement and relocation; (d) archaeological explorations; (e) community-based forest management; and (f) entry of the military.

In decision T-652-98 regarding the exploitation of natural resources in traditional territories of indigenous peoples, the Constitutional Court of Colombia argued that "…indigenous peoples are subjects of fundamental rights. If the State does not guarantee their right to subsistence (survival), these communities will not be able to materialize their right to cultural, social and economic integrity which is stated in the Constitution". Article 2 of the Constitution of Mexico (amended in 2001) recognizes the land rights of indigenous communities but subjects them to the rights of "third parties", a legal limitation which indigenous organizations and legal scholars consider rather as a step backwards in the recognition of their collective rights.

Indeed, the Special Rapporteur notes that numerous formally recognized legal rights of indigenous peoples are not fully implemented in practice, either in the courts by way of final adjudication determined by the judiciary, or as a result of new legislative acts which in fact weaken or reduce previously legislated rights. This concern has been expressed by indigenous participants at the WGIP. In relation to such regression in the case of Australia, the Committee on the Elimination of Racial Discrimination (CERD), recommended that "…close scrutiny continue to be given to any other proposed state and territory legislation to ensure that protection of the rights of indigenous peoples will not be further reduced".

In various UN and other forums, indigenous organizations have signaled their concern about negative impacts of major development projects on their environments, livelihoods, lifestyles and survival. One of the recurrent issues is the loss of land and territories that indigenous communities suffer. The lack of control over their natural resources has become a widespread worry. Very often these projects entail involuntary displacements and resettlement of indigenous communities which happen to lie in the way of a dam, an airport, a game reserve, a tourist resort, a mining operation, a pipeline or a major highway etc. As a result, violations of civil and political, economic, social and cultural rights occur with increasing frequency, prompting indigenous peoples to launch major protest or resistance campaigns in order to bring public attention to their plight, besides engaging the judicial system or appealing for administrative redress, as well as lobbying the political system.

A review of some recent complaints about alleged human rights violations of indigenous peoples in connection with activities surrounding the planning or execution of major development projects of different kinds draws attention to a number of focal points around the world. The High Court of Australia delivered a landmark decision on 8 August 2002, which denied native title rights over any mineral or petroleum resources in the Miriuwung-Gajerrong native title claim first lodged in 1994. A majority of the court found that native title rights did not apply to leases for the Argyle diamond mine or the Ord River irrigation project in Western Australia. The Mapuche people in Chile argue that they face the threat of physical and cultural disappearance caused by transnational logging companies. An indigenous community in Kenya reported to the UNWGIP that "today, this destruction of our cultures and land continues, due to so-called development projects such as mining, logging, oil exploration, privatization of our territories, and tourism". The Kickapoo Nation in Oklahoma, USA, is now struggling to maintain their very existence and the health of their land and water resources due to an impending superhighway from Canada to Mexico. It was reported that in Ecuador, oil activities are being undertaken which result in the break-up of the traditional, cultural and political structures of indigenous communities while facilitating the integration or assimilation of the oil economy in the country. In Japan, the building of a hydroelectric power dam in Nibutani, land sacred to the Ainu people, caused the destruction of traditional agriculture and the submergence of their sacred ceremonial sites. It further disrupted the links between the elders and the young as poverty forced families to sell their lands to the Government, which created divisions in the community.

Serious issues regarding the non-recognition of, and failure to respect, the rights of indigenous and tribal peoples have been reported in Suriname. Indigenous and tribal peoples (Maroons), who together comprise around 75,000 persons or about 14 % of the total population, occupy the forested areas of the 'interior' and suffer various types of discrimination in the national society. The government's report to the World Summit on Social Development recognizes these peoples as stakeholders in natural resources exploitation in their traditional lands but concedes that their participation in decision taking in those issues 'needs to be improved'. Legally, the land they occupy is owned by the state, which can issue land property grants to private owners. Indigenous and tribal lands, territories and resources are not recognized in law. Various indigenous and Maroon communities have been affected by mining (gold and bauxite) and logging activities carried out by national and foreign companies, without their prior consent or participation. As a result, numerous villages have had to relocate against their will and their environment has been disturbed, disrupting their traditional subsistence economy, their health, their social organization and their culture. Despite petitions to the national government and the Inter-American system of protection of human rights (Commission and Court), the indigenous and Maroon communities have not received the protection they require.

The Bakun Dam in Malaysia is reported to cause the forced displacement of 5,000–8,000 indigenous persons from 15 communities by clear-cutting 80,000 ha of rainforests. Indigenous peoples in Manipur, India were reported to suffer a similar fate caused by the building of 25 hydroelectric dams. Thousands of families of the Santhal Adivasi people in the Jharkhand province of India have reportedly been displaced as a result of the extraction of the minerals without proper compensation or economic security. In Thailand, several highland communities including the Karen people have reportedly been moved out of national parks against their will, whereas tourist development in Hawaii resulted in the displacement of indigenous people and their increasing poverty. Asian indigenous representatives expressed to the Working Group on Indigenous Populations that "…conflict and development interventions had resulted in large-scale displacements, internal and external, and serious consequences for [indigenous] children and youth resulted from the implementation of inappropriate and non-consultative development projects".

African indigenous peoples are not the exception when it comes to displacement from their traditionally owned lands. The creation of national parks or game reserves has forced people off their land. The Boran of Kenya, for instance, testified that four reserves created in Isiolo had been annexed affecting important grazing and watering points previously used by pastoralists. Moreover, the Keiyo indigenous people in Kenya also reported that they have been forcibly evicted from their land without compensation, because of mining activity there. Despite judicial appeal to the country's High Court (which was dismissed on technical grounds) and international concern, the Basarwa people in Botswana had their water supply cut off and have had no choice but to leave their traditional hunting grounds in the Central Kalahari Game Reserve for resettlement villages, to make way for government-sponsored development activities in the area.

Evictions or involuntary displacements are a common feature resulting from major development projects. The Committee on Economic, Social and Cultural Rights concluded that forced evictions are prima facie incompatible with the requirements of the International Covenant on Economic, Social and Cultural Rights. The term 'forced evictions' is defined as "the permanent or temporary removal against their will of individuals, families and/or communities from the homes and/or land which they occupy, without the provision of, and access to, appropriate forms of legal or other protection". Oftentimes, forced evictions occur in the name of development.

Conflicts over development projects on the lands of indigenous peoples lead to further violations of human rights. For instance, forced evictions from their traditional lands may lead to breaches of civil and political rights such as the right to life, the right to security of the person, the right to non-interference with privacy, family and home, and the right to the peaceful enjoyment of possessions. The Special Rapporteur has received reports about the arrest and harassment of indigenous persons involved in protest against destruction brought by the building of dams, and other extraction activities including logging and mining.

For example, people in Penan (Malaysia) have reportedly been arrested because they were blockading roads trying to stop loggers destroying their traditional forests. Philippine indigenous peoples have allegedly been physically abused and detained by mining companies and the police in the process of peaceful picketing against mining activities on their traditional lands. Sometimes, as in southern Africa, the strict enforcement of environmental conservation laws prevents indigenous farmers from farming their traditional land or using traditional resources, thus turning them into offenders who may be jailed for attempting to subsist. According to a recent report, oil workers in the Upper Pakiria River region of southeastern Peru forced the Kugapakori to move deep into the Amazon and threatened to arrest and decimate the community with diseases if they refused to leave their home. The Cucapá people in northern Mexico have been restrained by the authorities from practising their subsistence fishing because of environmental concerns, but the National Commission of Human Rights found that their human rights were being violated and recommended to the government in April 2002 that the Cucapá become participants in the planning and execution of programs for their own social development, including the fishing of protected species for their subsistence. Also in southeastern Mexico, indigenous squatters have been evicted from a biosphere reserve on environmental grounds, but NGOs refer to the various kinds of business interests wishing to invest in the area (see case study below on the Puebla Panama Plan).

Major development projects often entail serious health hazards for indigenous peoples. Environmental degradation, toxic chemical and mineral wastes, the destruction of self-sustaining eco-systems, the application of chemical fertilizers and pesticides are but some of the factors that seriously threaten the health of indigenous peoples in so-called 'development zones'. When relatively isolated indigenous communities enter into contact with the expanding national society and monetary economy—as has happened dramatically in the Amazon basin and other inter-tropical areas in recent decades—indigenous peoples also risk contracting contagious diseases, such as smallpox, aids and venereal diseases, as well as psychological troubles.

Indigenous peoples also argue that "environmental degradation and pollution [are] an integral facet of the health and well-being of indigenous peoples", citing, for instance, toxic contamination by *persistent organic pollutants* (POPs) and other industrially produced toxins. The Batwa in Rwanda report that deforestation of land leads to loss of traditional medicinal plants and to increased mortality. The right to food is also under siege by development projects, such as the construction of a dam in the Cuene region in Namibia which would significantly reduce or destroy food sources for the Epupa community by flooding the palm nuts and the faidberbia albida trees which provide a food supply for goats, a vital food source for the community. Because of the pollution of their traditional lands, the peoples of the north in Russia report that they have now become 'ecological refugees', whereas mining activities in Peru reportedly cause the pollution of fresh water used by indigenous peoples for food production. During the Special Rapporteur's mission to the Philippines in December 2002, numerous indigenous representatives reported similar environmental, economic and social effects of mining activities in various parts of the country, which they aptly label 'development aggression'.

Indigenous peoples have argued at length and legitimately that major development projects that do not take into account their fundamental interests entail violations of their basic human rights. At the UN Working Group on Indigenous Populations they maintain that "the indigenous approach to self-development [is] based on the principles of respect for and preservation of land, natural resources and all elements of the natural environment; consensus in decision-making; mutual respect for peoples' values and ideology, including sovereignty over land, resources and the environment under natural law". They also complain that full, meaningful and effective participation of indigenous peoples in development is generally not being considered. For instance, indigenous peoples from Chittagong Hill Tracts in Bangladesh said that "development strategies based on road construction, pacification programs and socio-economic development programs, and immigration, remained in the hands of the military and the participation of indigenous peoples in the development was excluded". The Ogiek of Kenya and the Batwa of Rwanda, referring to the need to get their views across, spoke of difficulties of ensuring effective minority participation in a majority-based democratic system.

On the other hand, some governments make efforts to ensure the participation of indigenous peoples in development. For instance, Canada adopted a number of initiatives in this direction such as participation of indigenous peoples in environmental assessment and regulatory boards and in land claim settlement agreements. It further developed a regional partnering approach to increase the opportunities for indigenous peoples' employment. New Zealand has launched the capacity building program designed to assist Whanau, Hapu, and Iwi Maori communities to identify needs and develop initiatives to achieve long-term economic development.

8.3 Selected Case Studies

Detailed research reports on major development projects and their impact on the lives and livelihoods of indigenous peoples as well as on the environment are available for a number of countries. A small selection of these experiences, particularly as regards the implications of the construction of major dams, are presented and summarized in the following sections.

8.3.1 Costa Rica

The Boruca hydro-electric project in southern Costa Rica, to become operational in 2012, is expected to flood an area of around 250 km^2 which would directly or indirectly affect seven indigenous territories and some non-indigenous areas as well. The Costa Rican Electricity Institute (ICE), which is promoting the project, has reportedly not formally consulted with indigenous organizations, which have organized commissions to dialogue with the government and have received help and advice from local universities and international non-governmental organizations. A technical study undertaken to assess the possible effects of the project on indigenous peoples draws attention to the expected displacement of the affected population, disruption of traditional agricultural activities, changes in the environment, disorganization of customary life in indigenous communities, short term employment for local people but no long-term plans for their incorporation into new economic activities, inflationary pressures on the cost of living and other worrisome consequences. The Special Rapporteur suggests that the government of Costa Rica would be well advised to promote mechanisms whereby the opinion of indigenous peoples may be taken into account in relation to the Boruca project.

8.3.2 Chile

During the 1990s important changes occurred in the Bio–Bio river basin in southern Chile, occupied by around 10,000 Mapuche-Pehuenches, due to a major hydroelectric development, involving eventually the construction of six different dams and electricity plants. The first of these, Pangue, built by ENDESA, a formerly public but now privatized company, was completed in 1996. Despite having government support and international financing, the company showed no regard for the needs and interests of the Pehuenche communities nor the local environment. An evaluation study commissioned by the World Bank, which had partially financed the project, was highly critical, pointing to the fact that the poor indigenous population in the area had not benefited at all from it, whereupon the distribution of the report to the Pehuenche people was withheld. A second study corroborated the earlier findings, prompting a statement by the Bank's president recognizing the mistakes and drawbacks of the project.

Nevertheless, the Chilean government and the corporation went ahead with plans to build the second, much larger, dam and plant at the Ralco site, to become operational in 2003. By that time, Chile had adopted new indigenous and environmental legislation, which enabled Mapuche organizations to challenge the projects politically as well as in court. The *National Corporation for Indigenous Development* (CONADI), a government agency, was charged with the task of negotiating an agreement between the parties, but two of its directors—both indigenous professionals—were sacked because they expressed their reservations about the way the company was handling the indigenous and environmental issues. Studies detailing the cumulative harmful effects of the six-dam project on the indigenous people and the environment, were rejected by the authorities. In fact, both CONADI and the National Environmental Agency (CONAMA) at one point advised the government to reject the project, but their positions were overruled. The Chilean government became concerned about mitigating the negative effects of the project on indigenous peoples, yet recognizes that the indigenous law is subordinate to other laws that in this case appeared to be paramount.

Despite the opposition of 4000 Pehuenches to their involuntary resettlement and the destruction of their traditional environment and way of life, and in complete disregard for the existing indigenous and environmental legislation (Chile has not yet ratified ILO Convention 169), the company (now part of a transnational corporation), continued to buy off individual Pehuenche families in exchange for their landholdings. By 2002 only seven families were holding out while the Ralco project was nearing completion. One of the problems for the Pehuenche is that their traditional collective landholdings and territories have been privatized by decree, making it easier for business interests to appropriate indigenous lands for their own purposes. A court ruled that the Pehuenches had priority when recovering land that was located above the water line.

Observers have noted that in the Ralco issue, business priorities, with state support, appear to override the social and environmental concerns that have been expressed by massive protests and court action undertaken by Mapuche organizations and their supporters. As the six-dam project on the Bio–Bio progresses, the future of the Pehuenche people, particularly the two local communities directly affected by the rising waters of the dam, Ralco-Lepoy and Quepuca-Ralco, looks bleak indeed and their traditional way of life appears to have been broken to the point of no return. Moreover, the Ralco case clearly shows the social tensions that arise between a 'modernizing' development model and the social, environmental and cultural costs to the people who bear the burden of this economic transformation. The government of Chile reports that indigenous peoples are not involved in the planning of major development projects, but once such projects have been decided upon, then indigenous communities may become involved in order to help mitigate possible negative effects of these projects. The Special Rapporteur suggests that Chile ratify ILO Convention 169 as soon as possible and that it abides strictly by emerging international standards and its own indigenous and environmental legislation in order to adequately protect the interests of indigenous peoples; indigenous communities should be involved directly, whenever major economic development projects that affect their lives and livelihoods are being considered.

8.3.3 Colombia

The Emberá-Katío indigenous people have traditionally lived in the area surrounding the Sinú and Verde Rivers in northwestern Colombia (departments of Córdoba and Antioquía). Their ancestral territories are legally recognized as two Indigenous Resguardos (Reserves), created in 1993 and 1996, and inhabited by about 500 families (ca. 2400 people). The Emberá-Katío are one of the several indigenous peoples who have suffered most from the persistent violence of Colombia's civil war. Over many years they have been negotiating with the authorities regarding the State's intention to allow a private company to build several large hydroelectric dams that would flood a good part (up to 7000 ha) of their traditional territories.

Concerned about the negative ecological and economic effects that the Urrá 1 dam would have on their cultures and social organization, the Emberá Katío traditional authorities (cabildos) have been subject to great pressures and been accused of being guerrilla supporters and 'enemies of progress'. Since 1992 some of their land was expropriated as being of 'public utility' and the privately owned Urrá company received a license to begin work on the project without prior consultation with the indigenous communities (mandatory according to the Colombian constitution).

In 1994 the company and Colombia's National Indigenous Organization (ONIC) agreed on a framework for mandatory consultation before the beginning of the second phase of the project, involving flooding and functioning of the dam. A proposed Ethno-Development Plan established compensation for eventual negative impacts of the dam on the Emberá-Katío. However as the river was diverted, new damaging impacts emerged, such as making it difficult for the indigenous to navigate and fish in the river. Despite an evolving conflict, the company obtained the government license to flood the area. This was later nullified by Colombia's Constitutional Court, which declared that the process violated the fundamental rights of indigenous peoples, and ordered a new consultation process as well as compensation for the Emberá Katío. In 1998 violence escalated, several indigenous families were forced to leave their homes under threat, property was destroyed, and more seriously, several indigenous leaders were assassinated or forcibly disappeared presumably by paramilitary forces, whereas others became the alleged victims of the *Revolutionary Armed Forces of Colombia* (FARC).

In 1999 the company was able to obtain another license for flooding, despite only partial consultation with the indigenous communities. Some of these refused to resettle notwithstanding the rising waters. Later in the year, a large delegation of Emberá Katío traveled to Bogotá, the country's capital, to protest against the situation, where they were put under intense political pressure. Finally, in 2000 a new agreement was reached between the government, the company and the indigenous communities. Besides promising social and health services to be provided by international agencies, the agreement acknowledged the Emberá Katio's neutrality, their full territorial autonomy, and their non-combatant condition. Nevertheless, violence continued against the Emberá in the form of assassinations, forced disappearances, arbitrary detentions and threats, some of which has been attributed to paramilitary groups and some to the FARC.

In June 2001 the *Inter-American Court of Human Rights* (IACHR) asked the government of Colombia to take 'urgent and concerted' measures regarding the disappearance of an Emberá leader, and to guarantee the right to life and the physical integrity of the rest of the community. It had to reiterate this appeal several days later as a result of government inaction. In 2002 further assassinations and forced disappearances decimated the Emberá Katío communities in the region. In October the Office of the United Nations High Commissioner for Human Rights in Bogotá issued a press statement denouncing the forced displacement of an

Emberá community of 800 people, including 250 children, due to threats by the FARC and called upon the national government to take adequate protective measures. In a letter to the Special Rapporteur, the ONIC restates its position that mega-projects are the main cause of current conflicts between the indigenous peoples and the State. As examples, the organization mentions the U'wa people and its ongoing conflict with Occidental Petroleum Co. (Oxy) over oil drilling on indigenous territory; the Emberá Katío and the Urra hydroelectric dam, the Wayúu and coal mining activities; another dam under construction in Saldaña where the Pijao people live; logging on Chamí forests by the Smurffit company; and the conflict between Inga, Kofane and Siona communities and oil companies over drilling and road building. More tensions are predicted among the Sikuani due to the channeling of the Meta river and an African palm plantation project, as well as the Emberá people in relation to the building of the proposed Inter-Oceanic Atrato-Truandó Canal.

The survival of the Embera-Katío people is at stake. Several of their most important and prominent leaders have been killed in the last five years. The Urrá I dam was proposed and is being built without their consent, involving involuntary displacements, social and economic disorganization and cultural disruption. They resent the construction of this dam as a threat to their way of life, and some of the impacts that have already been reported seem to support this view. These include: diseases which were unknown to the area, scarcity of fish and other basic elements of their diet, and most significantly, the disruption of the river, which represents a central place in their spiritual relationship of the Emberá Katío people to their land.

The situation of the Emberá Katío is not unique, because other indigenous peoples in the country face similar threats. Moreover, they have become, as other indigenous communities, victims of a violent civil conflict between armed parties involving the national security forces, the revolutionary guerrillas, the paramilitary groups as well as criminal elements linked to drug trafficking. They have proclaimed their autonomy and neutrality in these conflicts, demanding only that their territories, cultures and ways of life be respected. Unfortunately, this has not been the case and so their fundamental human rights have been and continue to be systematically violated. The Emberá Katío face the danger of not being able to survive this violence as a distinct people: a clear case of ethnocide.

8.3.4 India

The Sardar Sarovar Dam in India is the largest of 30 large, 135 medium and 3000 small dams to harness the waters of the Narmada river and its tributaries, in order to provide large amounts of water and electricity for the people of Gujarat, Maharashtra and Madhya Pradesh. With a proposed height of 136.5 m, the government claims that the multi-purpose *Sardar Sarovar Project* (SSP) will irrigate more than 1.8 million ha and quench the thirst of the drought prone areas of Kutch and Saurashtra in Gujarat. Others counter that these benefits are exaggerated and would never accrue to the extent suggested by the government. Instead the project would displace more than 320,000 people and affect the livelihood of thousands of others. Overall, due to related displacements by the canal system and other allied projects, at least one million people are expected to become uprooted or otherwise affected upon completion of the project. Indeed, the development surrounding the Narmada River has been labeled "India's greatest planned human and environmental disaster", a far cry from former Prime Minister Nehru's idealization of dams as the "secular temples of modern India".

Two-thirds of the over 40,000 families expected to be displaced by the reservoir's creation will be tribal people or Adivasis, belonging to different groups collectively referred to as Bhils. Displacement of Adivasis from their traditional lands and resources due to the creation of reservoirs, canals and reforestation projects significantly impacts on the ability of Adivasis to fully enjoy their human rights. They live mainly in 14 villages in Gujarat, 33 in Maharashtra and around 53 in Madhya Pradesh. The Adivasis are largely self-sufficient, growing their own food and collecting fuel, building materials, fodder, fruits, and other resources from the forests and common lands around their villages, as well as relying on water and fish from the river. Resettlement away from their territory means the destruction of their lifestyles and village organization. One farmer whose village will be submerged commented: "the forest is our moneylender and banker. From its teak and bamboo we built our homes. From its riches we are able to make our baskets and cots… From its trees we get our medicines".

In the early nineties opponents to the dam staged a series of non-violent protests (*dharnas* and *satyagraha*), prompting the World Bank, after commissioning an independent review which underlined the flaws in the project, to withdraw its remaining funding for it (the Bank cancelled $170 m remaining on its loan of $450 m). Work on the dam continued nevertheless, despite attempted judicial restraint, and by the summer of 2002 the water level in the reservoir rose much higher than initially expected, threatening many more people and villages with flooding. The government's rehabilitation and resettlement measures for 'oustees' (displaced persons) appeared to be insufficient, generating a number of protest activities by the affected villagers within the rising waters themselves. Protest against the project has remained strong and the Narmada Bachao Andolan (Save the Narmada movement) has been particularly instrumental in fostering awareness and dissent. Many activists and tribal people continue to maintain that they will never abandon their land to the dam, even if it means doobenge par hatenge nahi: death by drowning.

Multipurpose dams surely stimulate economic activity and have the potential for bringing benefits to large sectors of the population. The problem is whether these benefits are designed to reach the indigenous peoples who provide the land on which such projects are established, and how. It is estimated that the SSP will enable the irrigation of 1.8 million ha of land in Gujarat alone. Irrigation facilitates the production of food and other crops, which could significantly improve food production in drought prone areas. However, it appears that much of this area is unsuitable for irrigation because of water logging and salinization. Moreover, some of the designated water is likely to be consumed by sugar plantations before reaching more needy farms further away from the dam. Other potential benefits of the irrigation scheme and electric power generation from the dam are unlikely to benefit the Adivasi population.

Adivasis were not involved nor consulted in the dam construction process, on the premise that the project and the displacement of people was to serve a 'public purpose' which would provide a 'development opportunity' to the affected population. While some local governments did involve non-governmental organizations, an observer notes that "while NGOs can play an important supportive role they cannot substitute the voice of the affected people, nor can they replace what is the basic responsibility of the State".

Only the Adivasi population who live in the area that will be submerged in the reservoir (considered as *Project Affected Peoples* or PAP) are eligible for compensation and resettlement. However, many more will be affected indirectly, yet they are often not considered as PAP and therefore ineligible for rehabilitation. This would include areas affected by canals, rock-filled dykes, marooned islands, the creation of a new wildlife sanctuary and a reforestation scheme to compensate for tree loss and resettlement schemes on traditional Adivasi lands. Adivasi territory has also been affected by the construction of a colony to house the workers and officials engaged in the construction work and administration of the dam. All of these secondary consequences have displaced Adivasi villages and affected their lives and livelihoods. Patwardan comments that "displacement needs to be viewed as a 'process' rather than an 'event' which starts much before the actual physical displacement and continues for a long time after uprooting has taken place", and concludes that the current situation is symptomatic of the "gross underestimation of the human costs of large dams".

Whereas local state governments have offered comprehensive resettlement and compensation packages to 'landless' Adivasis displaced from their homes, observers point out that in practice Adivasis have not fully benefited from them. The promised lands in Gujarat did not materialize or were of poor quality, whereas in Madhya Pradesh the government had no resources to resettle displaced Adivasis. Moreover, resettlement has been delayed for many years and it is reported that 75 % of the displaced people have not been rehabilitated. To the extent that the law does not recognize customary rights to land and that therefore Adivasis may be considered 'encroachers' on government land, they have not received adequate compensation for their losses. In common with other indigenous peoples, Adivasis have a unique and close relationship with the land and its resources. Compensation packages treat land as property, whereas for Adivasis, their land is intrinsically linked to their culture and livelihood. It appears that the Government has omitted to deal with the numerous non-quantifiable losses experienced due to the dam such as loss of access to religious sites and social disintegration. Displacement due to the SSP has led to fragmentation of Adivasi communities as well as loss of cultural identity. Resettlement areas are often unsuited to the communal lifestyle of Adivasis, particularly if they have been resettled in communities of non-tribal people who reject the tribal way of life or have had to move to the cities.

Involuntary displacement readily leads to a violation of several economic, social and cultural rights. Despite claims to the contrary, resettled Adivasis have generally had to suffer a reduction in their standard of living, the loss of livelihood resources, and a reduction of health standards, a situation that stands counter to Articles 11 and 12 of the ICESCR. While in displaced communities government has established schools for the population, there are reports that due to economic hardship many children cannot afford to stay in school, whereas the curriculum appears to be ill adapted to the cultural and language needs of Adivasi children (Article 13). There have also been reports of violence and the use of force by the police upon protesters and resisters to displacement, in violation of ICCPR. The NBA recently called for protest of the Narmada Control Authority's decision in May 2002 to allow the dam height to rise to 95 m even though over 35,000 families displaced when the dam height reached 90 m have still not been resettled. In a recent urgent appeal to the Prime Minister of India the Habitat International Coalition reports that "Submergence due to the monsoons and raising the dam's height have destroyed the crops and homes of SSP-affected villages in Nandurbar District (Maharashtra) and Jhabua District (Madhya Pradesh), rendering the villagers homeless. These people now face a severe food and drinking-water shortage". It also reports that the Maharashtra government indicates an increase in the number of project-affected persons at the 95 m level, and admits that the government does not have enough land for rehabilitation of the affected persons.

The Sardar Sarovar dam and other similar projects on the Narmada river raise a number of complex issues. Originally, the interests and aspirations of the affected Adivasi population were not considered in the project design and implementation. As a result of continued lobbying by tribal and human rights organizations, the government of India now recognizes that the issues raised by the affected communities must be taken into account. Yet the implementation of measures intended to mitigate the negative effects and increase the benefits of the project for the Adivasi population has lagged behind and is considered as insufficient by the people involved. The Special Rapporteur recommends that the human rights of the Adivasis must be included as a foremost priority in the implementation of this development project and others of its kind. Only with the full and informed consent of the tribal people concerned will a truly human rights-centered development, as recommended by the General Assembly, become possible. An immediate step would be to halt any further rise in the reservoir's water level until the outstanding issues of rehabilitation and resettlement are fully solved to the satisfaction of the affected population, through constructive dialogue and negotiation between the parties. India could also signal its commitment to the human rights of its Adivasi population by ratifying ILO Convention 169 and approving the UN Draft Declaration on the rights of Indigenous Peoples. Alternative ways of involving the Adivasis in the project should also be considered. It has been suggested that they should be considered as partners in the project, with their investment being their natural resources. Adivasis qua investors would be entitled to share in the project's benefits.

8.3.5 Philippines

The San Roque Multipurpose Project in the Philippine Cordillera region involves the construction of a large dam on the Agno river which will be used primarily for power-generation and secondarily for irrigation and flood-control. Construction of the dam and power plant were completed in July 2002 and the water began to rise in August; operation of the power plant was scheduled to begin in January 2003. The construction site in the municipality of San Manuel, province of Pangasinan, covers about 34 km^2, but the irrigation and flood-control components will extend over a much wider area, involving around 30 municipalities in three provinces. The dam reservoir is expected to submerge eight small upland villages that are home to indigenous people.

Many other villages are bound to be affected by sediment build-up and upstream flooding as the reservoir gets silted. To mitigate the potentially negative impact of these processes, the implementation of a Lower Agno Watershed Management Plan is underway. The San Roque project is being implemented by the *San Roque Power Corporation* (SRPC) with credit financing from the *Japan Bank for International Cooperation* (JBIC). Several Philippine government agencies are actively involved in the project's implementation, particularly the watershed management, the irrigation and flood-control components. Whereas the power to be generated will range between 30 and 54 GW h monthly, the irrigation component of the project is aimed at extending, improving and integrating various existing irrigation works, so as to service more than 70,000 ha of riceland. The area to be serviced by the flood-control component is estimated at about 125,000 ha.

The area upstream of the dam is occupied by Ibaloy, Kankaney and Kalanguya indigenous peoples. About 120 households of eight indigenous villages have been dispersed by the local effects of the rising waters of the dam. Furthermore, nearly 5,000 indigenous households (about 26,000 individuals) are going to be affected by the sedimentation and flooding to be expected from the reservoir's eventual siltation, and more than 3,000 households will be affected by watershed management. A high rate of sedimentation takes place because of continued dumping of muck waste and impoundment of tailings from several large mining operations, which threatens to seriously alter the traditional activities of numerous indigenous communities in the area. The watershed management plan, intended to mitigate project-impact, involves curtailing some of the traditional activities of the indigenous communities, such as small-scale ore mining (which does little to affect the environment), banning the harvesting of timber products that are used for home-construction and kitchen-fuel purposes, and regulating subsistence swidden agriculture which is usually considered as sound agro-forestry management. Instead, large commerce-oriented agricultural production is being promoted as well as livestock raising for the market, that imply widespread clearing of vegetation and induced massive soil erosion in both the upper and parts of the lower river basin.

The project has several human rights implications. Firstly, environmental disruption; secondly, the displacement of population, some which appears to have been undertaken forcibly, but mostly through insistence on the implementation of the project in the face of community resistance and persuasion. Gradually, the people's resistance to the project has grown silent. Most importantly, indigenous people's land rights have been disregarded. Proprietary ancestral rights of indigenous families have not been given due recognition, but as project implementation progressed some families about to be displaced accepted some form of compensation, which was then cited as indication of consent. In fact, none of the affected communities participated in the planning of the project itself, and none freely gave their consent to its implementation. But many individuals participated in the consultations concerning impact-mitigation measures, and all of them are now bound by the enforcement of those measures, which imply drastic changes in livelihood engagements.

Whether deliberately or without meaning to the watershed managers are steering the households away from the peasant livelihood mix traditional to their indigenous communities, towards the monocultures that tend to define the production of vegetables, flowers, broom grass, and livestock for the market. Starting with their lending of capital for the new livelihood ventures, the watershed managers are introducing the households to new economic relations that may or may not be good for the communities. Whatever the final results, the debates stirred by the dam projects has already disrupted local social relations considerably.

This has occurred because local mechanisms for the protection of indigenous rights have not been effective. The indigenous communities of the municipality of Itogon tried to avail themselves of the mechanism provided by the Philippines' Local Government Code to withdraw endorsement of the dam, but the project continued. The Philippines' Indigenous Peoples' Rights Act provides for free and prior informed consent and enables an indigenous community to prevent the implementation of any project which affects its ancestral domain in any way by refusing consent to the project. Though Itogon's indigenous communities petitioned the National Commission on Indigenous Peoples (NCIP) to suspend the project because free and prior informed consent had not been given, the commissioners declined to act on the petition. Thus, the laws designed to protect the indigenous communities were in fact ignored.

8.4 Dams, Development and Human Rights

Lack of space does not allow the Special Rapporteur to report on other cases of large-scale development projects impacting upon indigenous peoples. The issues involving the construction of dams are, however, emblematic of the wider picture. Given their importance, the World Commission on Dams launched extensive studies on the matter, and concludes that:

> Large dams have significantly altered many of the world's river basins, with disruptive, lasting and usually involuntary impacts on the livelihoods and socio-cultural foundations of tens of millions of people living in these regions. The impacts of dam-building on people and livelihoods—both above and below dams—have been particularly devastating in Asia, Africa and Latin America, where existing river systems supported local economies and the cultural way of life of a large population containing diverse communities.

Concerning indigenous peoples specifically the report states:

> In the Philippines, almost all the larger dam schemes that have been built or proposed were on the land of the country's 6–7 million indigenous people. Similarly in India, 40–50 % of those displaced by development projects were tribal people, who account for just 8 % of the nation's one billion people. These costs are not balanced by any receipt of services from dams or by access to the benefits of ancillary services or indirect economic multipliers in the formal economy.
>
> ... For indigenous peoples and ethnic minorities dam-induced displacement can trigger a spiral of events that spreads beyond the submergence area. A case in point is the situation of the 100,000 Chakma people displaced by the Kaptai hydropower dam in the Chittagong Hill Tracts, Bangladesh. The project submerged two-fifths of their cultivable land; as a consequence, 40,000 Chakma left for India and another 20,000 were supposed to have moved into Arakan in Burma.
>
> ...The Bayano dam in Panama that forced the indigenous Kuna and Emberá peoples from their traditional territories resettled them on land that was less fertile and subject to encroachment by loggers. The Panamanian government systematically failed to fulfill agreements made with the affected indigenous people at the time of construction, as well as commitments negotiated later. Among the violations was the government's failure to compensate adequately for the loss of traditional territories and provide legal titles to the new lands. What happened in Panama in the 1970s is similar to what has happened in Malaysia in the 1990s. In the case of the Bakun project, rights to indigenous common land in the Ulu Belaga site were not recognised or properly assessed. Industrial countries' experience with indigenous peoples in the era of building large dams was not very different from that of developing countries. Dams built during the 1950s and 1960s cost the indigenous nations of the Missouri River basin in the United States an estimated 142,000 ha of their best land, including a number of burial and other sacred sites, leading to further impoverishment and severe cultural and emotional trauma. A guarantee used to rationalize the plan—that some 87,000 ha of Indian land would be irrigated—was scrapped as the project neared completion.

Another case is the second stage of the Churchill Rivers project in Labrador, Canada, consisting of two dams and two river diversions that will flood a large area of hunting territory of the Innu people who live on both sides of the provincial boundary. The Innu have yet to be clearly recognized as the owners of their lands, and the whole area is the subject of an unresolved Innu land claim currently being negotiated with the Canadian government.

The Commission recommends that in the future major development projects such as dams be approached on the basis of the recognition of rights and the assessment of risks, which is of particular relevance to indigenous peoples.

> The recognition of rights and the assessment of risk identify the interested and affected parties who possess rights or entitlements as well as risk takers and bearers. This opens the way for a negotiated approach that enables the decision-making process to assess options and reach project agreements. Those whose rights are most affected, or whose entitlements are most threatened, have the greatest stake in the decisions that are taken. The same applies to risk: those groups facing the greatest risk from the development have the greatest stake in the decisions and, therefore, must have a corresponding place at the negotiating table.

> Further, the Commission has sought to demonstrate that an approach based on the recognition of rights and assessment of risks can lay the basis for greatly improved and significantly more legitimate decision-making on water and energy development. This is an effective way to determine who has a legitimate place at the negotiation table and what issues need to be included on the agenda.

The debate on dams and indigenous peoples has wider implications, as reflected in the discussions on the environment and sustainable development. The *UN Conference on Environment and Development* (UNCED 1992) recognized that: "Indigenous people... have a vital role in environmental management and development because of their knowledge and traditional practices. States should recognize and duly support their identity, culture and interests and enable their effective participation in the achievement of sustainable development" (Rio Principle 22). Ten years later, the *World Summit on Sustainable Development* took a small step further by reaffirming: "...the vital role of the indigenous peoples in sustainable development". These statements must necessarily be taken into consideration seriously in the design, planning and execution of major development projects that affect the lives and livelihoods of indigenous peoples.

These issues stand out clearly in the ambitious *Puebla-Panama Plan* (PPP) adopted by the governments of Central America countries and Mexico in 2000, designed to modernize and integrate the region which shares a number of common features, including a high density of indigenous inhabitants and generally low levels of human development. The plan is designed to promote economic development through public and private investments, with international financing. A number of planned projects, which range from Airport Security to Fiber Optics Networks, organized around eight distinct regional initiatives, directly involve indigenous peoples, such as highway construction, tourist promotion, natural resource management, the introduction of new crops and the setting up of maquila plants. Many indigenous and human rights organizations in the region have expressed their serious concern regarding the possible negative effects that a number of these projects, taken together, may have on indigenous human rights. Whereas the *Inter-American Development Bank* (IADB) expects the PPP to "take advantage of the human and ecological riches of the Mesoamerican region within a framework of sustainable development and respect for its ethnic and cultural diversity", the *Central American Indigenous Council* (CICA) addressed a letter to the region's presidents stating:

> 2. The indigenous peoples express to you our concerns about the absence and lack of enforcement of judicial and economic mechanisms to protect the territorial security of our peoples; and we state the need for the inclusion in Plan Puebla Panama of a regional strategy that guarantees that territorial security.
>
> We exhort the Nation States to create national judicial instruments to ratify and enforce the international instruments that protect indigenous peoples' rights. We urge the Presidents to frame the strategic actions of the Plan Puebla Panama in the promotion, guarantee and development of the indigenous peoples' fundamental rights, contemplated in the afore-mentioned instruments.
>
> It is imperative to create an indigenous component of the Plan Puebla Panama to facilitate the exercise of a transversal approach among the different components of the general strategy and to strengthen indigenous peoples' initiatives oriented to promote development with identity, equity and social justice.

Some indigenous organizations are more critical of the Plan. Several human rights organizations in the Isthmus of Tehuantepec, Mexico, are concerned that the implementation of the Plan in their region will destroy their traditional environment and natural resources, impact negatively on their subsistence agricultural activities and social organization and force them to accept low paying jobs in export-oriented assembly plants (*maquiladoras*). They specifically oppose the construction of a highway that would cross their traditional habitat without bringing them any benefits, and complain that their concerns have not been addressed by the agencies involved in promoting the PPP.

The Special Rapporteur wishes to transmit to the Commission on Human Rights his concern that notwithstanding statements to the contrary by the highest authorities and the various national and international agencies involved in promoting the Puebla Panama Plan as a high-priority project for regional integration and development, there are as yet no institutional and legal mechanisms in place for the effective protection of the human rights of the indigenous peoples of the area designed to offset the potential risks and threats to these peoples that the implementation of the Plan implies, nor are there as yet any effective mechanisms to ensure the full and informed participation of these peoples in the design, planning, execution and evaluation of the numerous specific projects foreseen in the Plan that may have considerable impact on the region's indigenous communities. He calls on the international financing agencies, the international and national business community and the region's governments to attach the highest priorities to the needs and concerns of the indigenous peoples in this matter, recalls the principles of a human-rights centered development approach and calls their attention to the declaration of the World Summit on Sustainable Development about the vital role of the indigenous peoples in sustainable development.

8.5 Conclusions and Recommendations

The issue of extractive resource development and human rights involves a relationship between indigenous peoples, governments and the private sector, which must be based on the full recognition of indigenous peoples' rights to their lands, territories and natural resources, which in turn implies the exercise of their right to self-determination. Sustainable development is essential for the survival and future of indigenous peoples, whose right to development means the right to determine their own pace of change, consistent with their own vision of development, including their right to say no. Free, prior, informed consent is essential for the human rights of indigenous peoples in relation to major development projects, and this should involve ensuring mutually acceptable benefit sharing, and mutually acceptable independent mechanisms for resolving disputes between the parties involved, including the private sector.

To the extent that international financial institutions such as the World Bank play a vital role in facilitating major development projects by providing various forms of financial support, the current revision of the World Bank's policy regarding indigenous peoples is of major importance. The WB has a specific policy on indigenous peoples designed to ensure that "World Bank-financed development projects do not have adverse impacts on indigenous peoples, and that project benefits are tailored to the specific needs of indigenous peoples". But some indigenous consultants argue that "the draft policy fails to uphold international human rights standards applicable to indigenous peoples..." and insist particularly on the collective rights of indigenous peoples to their customary land and territories. The Special Rapporteur recommends that the new Bank Policy for indigenous peoples should strictly adhere to all existing and evolving international indigenous human rights standards.

Any single major development project in indigenous areas may have either, or both, positive and negative effects directly or indirectly on indigenous peoples and communities. Under pressure from governments and NGOs, some business enterprises may undertake special efforts to improve the management of the surrounding environment, as well as to provide compensation, employment and/or social services (such as housing, schools, medical care, utilities) to the affected communities. Unfortunately, as we learn from indigenous organizations and research reports, these companies appear to be in the minority. Others, when faced with social protest and political opposition, or considering the cost of becoming involved in sustainable and human rights-centered development, prefer to close down their operations, withdraw their projects or abstain from making their investments. Still others, however, make use of different kinds of pressure (including violence or the threat of violence) to carry out their operations despite opposition.

Whereas human rights violations occurring in isolated cases may be dealt with successfully or not by the affected communities on an ad hoc basis, it is rather the long-term effects of a certain pattern of development that entails major violations of the collective cultural, social, environmental and economic rights of indigenous peoples. Within the framework of the globalised market economy the traditional environment becomes altered irreparably, non-renewable natural resources are destroyed and extracted exclusively for private gain, numerous communities and masses of people are uprooted, evicted or resettled with little or no regard to their actual needs and rights, sometimes accompanied by organized violence intended to intimidate, harass and make them comply with decisions taken by outside interests without or explicitly against their consent. Often, the same results are achieved through bribery, corruption and cooptation.

Whilst indigenous peoples have made important advances in recent decades, they are still considered in many countries as secondary citizens whose needs and aspirations are seldom taken into account by the powers that be. They are often denied effective political participation in government and the electoral system, and their concerns are hardly being met by established political parties. Nor have local and national power structures been favorable to the empowerment of indigenous peoples. If their human rights are to be effectively protected, they must be able to participate freely as equal partners and citizens in the decision- making processes that affect their future survival as specific peoples. This also means that their voices must be heard and their demands and grievances be met when major decisions are taken at the national and international level regarding development priorities and the allocation of resources. This is not yet the case, and the Special Rapporteur hopes that the Permanent Forum on Indigenous Issues will be able to fill this void.

The human rights of indigenous peoples and communities must be considered of the utmost priority when development projects are undertaken in indigenous areas. Governments should take the human rights of indigenous peoples as a crucial factor when considering the objectives, costs and benefits of any development project in such areas, particularly when major private or public investments are intended.

Potential investors must be made aware at all times that the human rights of indigenous peoples should be a prime objective when investment decisions in development projects are made in such areas or are expected to affect indigenous peoples directly or indirectly. There can be no justification for ignoring them.

Sustainable development must be understood not only in terms of environmental management but also as respectful of human rights at all times, particularly of indigenous peoples. Any development projects or long-term strategy affecting indigenous areas must involve the indigenous communities as stakeholders, beneficiaries and full participants, whenever possible, in the design, execution and evaluation stages. The free, informed and prior consent as well as the right to self-determination of indigenous communities and peoples, must be considered as a necessary precondition for such strategies and projects. Governments should be prepared to work closely with indigenous peoples and organizations to seek consensus on development strategies and projects, and set up adequate institutional mechanisms to handle these issues.

Potential long term economic, social and cultural effects of major development projects on the livelihood, identity, social organization and well-being of indigenous communities must be included in the assessment of their expected outcomes, and must be closely monitored on an ongoing basis. This would include health, nutrition, migrations and resettlement, changes in economic activities, levels of living, as well as cultural transformations and socio-psychological conditions, with special attention given to women and children.

To the extent that major development projects impinge upon traditional indigenous territories or ancestral domains, indigenous land and property rights must be considered as human rights at all times, whether they are so recognized legally or not.

Indigenous organizations should attempt to present their viewpoints publicly on major developments at an early stage and be prepared to work with governments, multilateral financing institutions and private companies to find convenient solutions to contentious issues. Non-governmental organizations are urged to support such efforts, particularly as regards the possibility of preparing and promoting alternative development strategies and projects within a human rights-centered approach.

Contentious issues between indigenous peoples, governments and business enterprises, arising in the course of the implementation of major development projects, should at all times be considered within the framework of democratic governance, open dialogue and negotiations, and should never be handled primarily as a problem of national security, law and order which often leads to military or police action that may violate the human rights of indigenous communities.

International organizations such as development banks and UN agencies in the field, should at all times be ready to support indigenous peoples and communities in making human rights the primary focus of development cooperation involving major development projects in indigenous areas.

The Special Rapporteur noted a recommendation of the 2001 UN Workshop on *Indigenous Peoples, Private Sector Natural Resource, Energy and Mining Companies and Human Rights*, which requested the OHCHR to continue to act as a facilitator for dialogue among indigenous peoples, Governments, and the private sector with regard to the issue of indigenous peoples' human rights and the private sector. In this regard, the Special Rapporteur endorses this recommendation, and further encourages the OHCHR to organize a second workshop on the topic along with appropriate human rights training for representatives of companies on international indigenous human rights.

Stavenhagen with displaced indigenous Bedouins in Naqab (Negev), Israel, 2009

Chapter 9
Study Regarding the Best Practices to Implement the Recommendations of the Special Rapporteur (2007)

The United Nations and its member states do not only wish to be informed about human rights abuses, but they are also keen on learning from different experiences concerning solutions to such violations and from policies undertaken by governments and public and private institutions, as well as by indigenous communities and organizations themselves that may lead to conflict resolution and the improvement of the human rights situation of these peoples. In this report to the Human Rights Council, the last one of my mandate as Special Rapporteur, I look at what in UN parlance is known as 'best practices'. (This report has been circulated by the UN as document A/HRC/4/32/Add.4 and is in the public domain.).

Abstract The study presents a number of general considerations concerning the objectives and impact of the Special Rapporteur's report, and provides specific examples of initiatives undertaken in specific countries to follow up on the Special Rapporteur's recommendations that have involved international organizations and agencies, civil society and indigenous peoples, in cooperation with the Governments concerned. The final part of the study incorporates a number of examples concerning specific countries in which these recommendations have promoted specific changes in State policies and legislation. The study concludes that, while the Special Rapporteur's reports have had an important impact in some countries, the recommendations incorporated in his reports do not generate automatic and speedy changes in the situation of the rights of indigenous peoples. The several initiatives that have been undertaken over the last years by Governments, the United Nations system, civil society and indigenous organizations to monitor and promote the implementation of these recommendations demonstrate that, if left for institutional action alone, the recommendations are rarely implemented. Implementation needs to be pushed forward in close cooperation with the Government and other stakeholders, including indigenous peoples themselves. In countries where follow-up mechanisms exist, institutional efforts for implementation have been more sustained, leading to concrete changes in law and practice. These experiences suggest that, despite the advances that can be identified, the general record of implementation of the Special Rapporteur's recommendations is gloomy. Much remains to be done by Governments, international agencies and other relevant stakeholders to bridge the 'implementation gap' that divides international and domestic norms and the serious human rights violations that indigenous peoples continue to experience in all parts of the world. The study contains a number of conclusions and recommendations to enhance implementation.

R. Stavenhagen, *Peasants, Culture and Indigenous Peoples*,
Texts and Protocols 4, DOI: 10.1007/978-3-642-34153-3_9,

9.1 Introduction

In resolution 2005/51, the Commission on Human Rights requested the Special Rapporteur on the situation of human rights and fundamental freedoms of indigenous people, Rodolfo Stavenhagen, to begin preparing a study regarding "best practices carried out to implement the recommendations contained in his general and country reports" (para. 9) and to submit a progress report to the Commission at its sixty-second session and the final study at its sixty-third session.

Following this request, the Special Rapporteur presented a progress report (E/CN.4/2006/78/Add.4) to the first session of the Human Rights Council in September 2006 containing an overview of the main conclusions and recommendations from his thematic and country reports; a summary of the information received from Governments, international agencies and civil society organizations on the actions being taken; and a plan of work for the preparation of the final study.

The Special Rapporteur would like to note that an in-depth study would have required full-time research and additional information. In this context, the present report should be seen by the Council as a general overview of the actions being taken and the challenges ahead that could serve as a first step for a more comprehensive study on the subject matter in the future.

Commission on Human Rights resolution 2001/57 establishing the mandate on the situation of the human rights and fundamental freedoms of indigenous people attributes to the Special Rapporteur the responsibility of formulating "recommendations and proposals on appropriate measures and activities to prevent and remedy violations of the human rights and fundamental freedoms of indigenous people" (para. 1 (b)). Such recommendations are included in a number of thematic and country reports. Since his appointment in 2001, the Special Rapporteur has presented six annual reports. In the first, the Special Rapporteur proposed a list of issues on which he wanted to focus his subsequent reports (E/CN.4/2002/97, para. 113), which was endorsed by the Commission (resolution 2002/65, para. 5). Subsequently, the Special Rapporteur prepared thematic reports on the impact of large-scale development projects (E/CN.4/2003/90); access to the administration of justice and indigenous customary law (E/CN.4/2004/80); education (E/CN.4/2005/88); and the implementation of legislation and jurisprudence concerning the rights of indigenous peoples (E/CN.4/2006/78). The Special Rapporteur presents his sixth annual report at the present session of the Council (A/HRC/4/32), which focuses on the state and evolution of the rights of indigenous peoples in recent years.

The Special Rapporteur has also submitted reports on his missions to Guatemala (E/CN.4/2003/90/Add.2); Philippines (E/CN.4/2003/90/Add.3); Mexico (E/CN.4/2004/80/Add.2); Chile (E/CN.4/2004/80/Add.3); Colombia (E/CN.4/2005/88/Add.2); Canada (E/CN.4/2005/88/Add.3 and Corr.1); South Africa (E/CN.4/2006/78/Add.2); New Zealand (E/CN.4/2006/78/Add.3). At the current session of the Council, the Special Rapporteur presents reports on his missions to Ecuador (A/HRC/4/32/Add.2) and Kenya (A/HRC/4/32/Add.3).

In preparing his final study, the Special Rapporteur used the information included in the replies to a questionnaire distributed in October 2005 which he received from the Governments of Argentina, Belarus, Canada, Chile, Denmark, El Salvador, Estonia, Finland, Germany, Lebanon, Mexico, the Philippines, the Russian Federation, Switzerland and Tunisia. The Special Rapporteur received replies from the *Food and Agriculture Organization of the United Nations* (FAO), the *United Nations Educational, Scientific and Cultural Organization* (UNESCO), the *World Health Organization* (WHO), the *World Food Programme* (WFP), the *Office of the United Nations High Commissioner for Refugees* (UNHCR), the *United Nations Development Programme* (UNDP), the *United Nations Population Fund* (UNFPA), the *United Nations Institute for Training and Research* (UNITAR), the *International Labour Organization* (ILO), the World Bank, as well as the country offices of the Office of the *United Nations High Commissioner for Human Rights* (OHCHR) in Colombia, Guatemala and Mexico, and the *OHCHR Regional Office for Latin America and the Caribbean*, in response to another specific questionnaire addressed to the United Nations agencies and programs.

This study is also based on the information compiled during the Special Rapporteur's participation in a number of visits, seminars and meetings, including the International expert seminar on the implementation of the Special Rapporteur's recommendations, organized by Rights and Democracy in Montreal, Canada, in October 2006. The Special Rapporteur received written contributions from a number of indigenous organizations, NGOs and individual experts. He acknowledges the cooperation received and wishes to thank all the people and organizations that supported this research.

The study first presents a number of general considerations concerning the objectives and impact of the Special Rapporteur's report, and makes a number of preliminary conceptual clarifications concerning the scope of the study. The second part of the study provides a number of examples of initiatives led by international organizations and agencies, civil society and indigenous peoples to follow up on the recommendations of the Special Rapporteur's reports, in cooperation with the Governments concerned. The third part analyses a number of instances in which these recommendations have promoted specific changes in State policies and legislation. The study concludes with a number of conclusions and recommendations to enhance implementation.

9.2 General Considerations on the Objectives and Scope of the Special Rapporteur's Recommendations

In its resolution 2005/51 the Commission on Human Rights specifically limited the scope of the study to the recommendations contained in the Special Rapporteur's 'general and country reports'. The emphasis on 'best practices' is particularly relevant in order to ascertain the effectiveness of the Special Rapporteur's mandate and the cooperation of the relevant stakeholders, particularly States, with this special procedure.

The 'best practices' approach presents methodological limitations related to the difficulty of establishing clear relations of causality between the Special Rapporteur's recommendations and policy and practical changes that have actually taken place. The Special Rapporteur's work is informed by and builds upon existing international standards regarding indigenous rights, including treaties, customary law and 'soft law'; the decisions and recommendations of international human rights bodies responsible for monitoring those norms, which have developed a specific jurisprudence concerning indigenous peoples; and other special procedures of the Human Rights Council (see E/CN.4/2002/97, paras. 6–33, and E/CN.4/2006/78, paras. 7–13, 51–79). Therefore, the recommendations made by the Special Rapporteur cannot be seen in isolation, but are rather part of the wider system of international norms, actors and procedures that interact to promote the rights of indigenous peoples.

Examples of this interaction are manifold. The Special Rapporteur's thematic reports have been used as a source in the reports of the Inter-American Commission on Human Rights and also in the activities of the *Working Group on Indigenous Populations/Communities of the African Commission on Human and Peoples' Rights*. His reports have also been used in the work of other special procedures of the Human Rights Council. For instance, the thematic report on the impact of major development projects is a tool for ongoing discussions within OHCHR concerning the impact of business on human rights, and for the work of the Special Representative of the Secretary-General on the Issue of Human Rights and Transnational Corporations and other Business Enterprises. In addition, the Special Rapporteur's country reports have been used by the United Nations treaty bodies in the preparation of their concluding observations concerning State compliance with the human rights conventions they have ratified.

Similarly, the Special Rapporteur's recommendations are related to social, political and legal processes at the domestic level. The different issues highlighted by the Special Rapporteur, particularly in his country reports, are derived from his independent assessment of already existing discussions and demands concerning the rights of indigenous peoples in the countries he visits. As a consequence, the implementation of the Special Rapporteur's recommendations cannot generally be seen in isolation from ongoing efforts by government actors, civil society organizations and indigenous peoples themselves to promote solutions to the substantive human rights issues that the recommendations seek to address.

The human rights situation of indigenous peoples is derived from complex historical processes and structural phenomena, and therefore the actions and strategies required to improve this situation are necessarily multifaceted. In a number of cases, the effective protection of indigenous rights requires specific legal, institutional and even constitutional reforms to guarantee them or to solve conflicts with other existing norms at the domestic level, and the implementation of these recommendations may be relatively easy to assess. In other instances, particularly when addressing broader or systemic conditions affecting the enjoyment of basic human rights by indigenous peoples, the Special Rapporteur's recommendations are phrased differently. The implementation of the recommendations must be measurable, and a system of benchmarks should be set to evaluate progress, with the participation of indigenous peoples themselves.

The impact of the Special Rapporteur's work on the protection of the rights of indigenous peoples is not measured necessarily only along the implementation/non-implementation continuum. His missions in several countries and the specific recommendations in his country reports have in some cases had a direct impact. Some of the participants in the Montreal expert seminar pointed out that the Special Rapporteur's country visits and reports possibly constitute one of the more effective, practically oriented lines of action of the various activities undertaken within his mandate.

Specifically, indigenous peoples themselves become involved in the visits of the Special Rapporteur. Typically, he holds consultations with indigenous organizations and individuals at the national, regional and community levels. These meetings have provided him not only with valuable information, but have also promoted a space for dialogue between indigenous peoples, Governments and other actors at the national level. In New Zealand, the visit was reportedly seen as a basic point of reference by indigenous organizations, irrespective of the level of implementation of the specific recommendations by the Government. The visit by the Special Rapporteur to Colombia was also seen by indigenous organizations as a crucial event for their empowerment. An expert at the Montreal seminar pointed out that the visit encouraged the consolidation of a distinct human rights agenda for indigenous peoples, and helped reinforce the relationships with human rights NGOs.

Though not on official mission, the Special Rapporteur visited Norway twice during his mandate at the invitation of the Saami Parliament and the University of Tromsø. In 2006, after lengthy negotiations, the Parliament adopted the Finnmark Act, a new law regarding the management of the Saami traditional reindeer-herding areas in the north of the country. The Special Rapporteur has been informed both by government officials and Saami spokespersons that his presence in the country during crucial stages in the process was considered a positive contribution to the adoption of the law.

The relatively high impact of country reports in public debates and policy-making concerning the rights of indigenous peoples at the national level, as well as the concrete character of some of the recommendations allow for a detailed analysis of their follow-up by the Governments and other actors concerned. Indeed, as this study shows, the most relevant 'best practices' in the implementation of the Special Rapporteur's recommendations relate to those in the various country reports.

One of main conclusions of the Montreal expert seminar was that the implementation of recommendations included in the Special Rapporteur's thematic reports has been limited in comparison to those in the country reports. This is partly due to their different objectives. Thematic reports aim at providing an overview of evolving domestic and international legal norms and policies, as well as the major challenges regarding the rights of indigenous peoples, with a view to calling international attention to areas of special concern. Their recommendations are not addressed to specific States, and government institutions do not often feel directly concerned about their implementation. It has been pointed out, however, that the Special Rapporteur's thematic reports are increasingly seen as authoritative sources for different purposes at the national and international levels. For instance, the Special Rapporteur's recommendations have served as a tool in the formulation of national policies, such as in the case of the Spanish Strategy of Cooperation with Indigenous Peoples (*Estrategia de la Cooperación Española con los Pueblos Indígenas*, ECEPI), to which the Special Rapporteur was requested to give an input.

Finally, while the 'best practices' study commissioned by the Commission on Human Rights constitutes a useful tool to assess the impact and effectiveness of the Special Rapporteur's recommendations, he cannot conclude these general considerations without noting that, as described in the thematic report presented to the current session of the Human Rights Council, despite the many efforts deployed, indigenous peoples around the world continue to suffer serious and systematic violations of their rights, a situation that will persist as long as the root causes of these violations remain unaddressed. In many cases, instead of 'best practices', the Special Rapporteur finds only 'good intentions'.

9.3 Follow-Up of Recommendations

In a number of countries, specific initiatives have taken place to follow up on the Special Rapporteur's recommendations. These initiatives have involved international organizations and agencies, civil society and indigenous peoples, in cooperation with the Governments concerned. These initiatives have been key in promoting 'best practices' in the implementation of the Special Rapporteur's recommendations in the countries concerned, and provide positive examples that could be applied to other countries.

9.3.1 OHCHR Project in Mexico and Guatemala

In 2005, the United Nations High Commissioner for Human Rights' (OHCHR) country offices in Mexico and Guatemala, in cooperation with the respective Governments, initiated the project Promotion and protection of human rights of indigenous peoples in Central America with special focus on Guatemala and Mexico. One of the main objectives of this project is to provide support to both Governments in implementing the recommendations of the Special Rapporteur's country reports, particularly by setting up human rights protection and monitoring standards to measure the implementation of the recommendations, the developments in the legal system, and the changes in the human rights situation of indigenous peoples and of women in particular.

In the framework of this project, OHCHR has promoted training courses for members of the Government, the judiciary and indigenous organizations on the rights of indigenous peoples. The project also promoted the dissemination of the reports by way of printed and audio materials in Spanish and indigenous languages. In 2006 two research projects on the recognition of traditional indigenous law in the official legal system were initiated in Mexico, following up the Special Rapporteur's recommendations on indigenous law and access to justice, and on the situation of the rights of indigenous women.

OHCHR Mexico and its counterparts in the Government have organized a number of meetings to evaluate the state of implementation of his recommendations, including one with high-level government officials in 2006, and a national consultation with indigenous and human rights organizations in January 2007. The project also supported the follow-up visit undertaken by the Special Rapporteur to the 'La Parota' hydroelectric project and other indigenous communities in the State of Guerrero in August 2006.

Similar meetings have taken place in Guatemala, where, at the invitation of the Government, the Special Rapporteur conducted a follow-up mission in May 2006. During his visit, he met with the President's full Cabinet, as well as with several governmental agencies and committees; members of parliament and the judiciary; indigenous and civil society organizations and representatives of the *United Nations Country Team* (UNCT). He further participated in a national workshop with more than 100 representatives of indigenous and civil society organizations, which presented him with a full assessment of the state of implementation of the recommendations of his country report.

In 2006 OHCHR Mexico conducted a survey on actions taken by government institutions, the legislative and judicial branches, as well as national human rights institutions at the federal and state levels to implement the Special Rapporteur's recommendations concerning that country. This information has been submitted to the Special Rapporteur and will also be presented in meetings with government officials. In Guatemala, the Office has assisted the Presidential Commission on Human Rights (*Comisión Presidencial de los Derechos Humanos*, COPREDH) in the elaboration of indicators to improve monitoring of the Special Rapporteur's recommendations.

The OHCHR binational project has also helped further the action of OHCHR country offices in the field of indigenous rights in those two countries. In Mexico, the Office identified the administration of justice in the State of Oaxaca as one of the priority areas for 2005. In planning the different activities in this area, consideration was given to the Special Rapporteur's recommendations in his report on administration of justice and indigenous law.

9.3.2 Other OHCHR Projects

Following the example of the project in Mexico and Guatemala, OHCHR launched the 'Andean Project', in 2006, aiming at working with the Governments of Bolivia, Ecuador and Peru in reinforcing the existing protection of the rights of indigenous peoples and mainstreaming indigenous issues in the work of the UNCTs. One of the lines of work of the project is the implementation of recommendations by United Nations treaty bodies and special procedures as regards the rights of indigenous peoples, including the Special Rapporteur.

In 2006, the OHCHR Andean Project, the UNICEF Regional Office and the United Nations Development Fund for Women Andean Regional Office started a study on the best practices and obstacles regarding the implementation of the Special Rapporteur's thematic recommendations in Ecuador, Bolivia, and Peru.[1] The study will pay special attention to the recommendations concerning indigenous children and women, in connection with the recommendations to these countries of the Committee on the Rights of the Child and the Committee on the Elimination of Discrimination against Women. The study, which is expected to be concluded in 2007, intends to promote the mainstreaming of the Special Rapporteur's thematic recommendations in policymaking and United Nations programming, including concerning the Millennium Development Goals.

In Ecuador, the Andean Project has led the first efforts to establish a follow-up mechanism to the Special Rapporteur's report on the visit to that country in April/May 2006. These efforts involve indigenous organizations through the *Permanent Advisory and Consultative Council of the United Nations and Organizations, Nationalities and Indigenous Peoples* of Ecuador The Council was established in the context of the *Human Rights Strengthening* (HURIST) program, a joint initiative implemented at country level by OHCHR and UNDP that endeavors to mainstream human rights in the work of the UNCT. One of the first initiatives undertaken by the Andean Project was the dissemination of the information concerning the Special Rapporteur's mandate and activities.

In his report on Colombia, the Special Rapporteur signaled the existence of serious conflicts as a result of faulty consultation processes in development projects in indigenous *resguardos* (reserves), and called upon the Government to work out "[a]n agreed approach to the consultation process" (E/CN.4/2005/88/Add.2, para. 108). OHCHR Colombia is currently considering the establishment of a specific program on promoting the right to consultation which would engage indigenous and Afro-descendant communities, government ministries and agencies, and the Office of the Ombudsman.

In the report on his visit to Chile, the Special Rapporteur recommended that OHCHR should organize a follow-up meeting "to identify ways in which the United Nations system can assist the State authorities in implementing the

[1] See OACNUDH-Comité Andino de Servicios, Mandato del Relator Especial sobre la situación de los derechos humanos y las libertades fundamentales de los indígenas, Lima, 2006.

recommendations set out in this report" (E/CN.4/2004/80/Add.3, para. 82). Since the report was made public in 2004, indigenous organizations have approached the Office on several occasions to seek its support in advancing the Special Rapporteur's recommendations,[2] 2 and the OHCHR Regional Office for Latin America and the Caribbean participated in various activities aimed at the dissemination and follow-up of the Special Rapporteur's recommendations. In 2006, the OHCHR Regional Office included these objectives as part of the Action 2 Project on strengthening the capacities of UNCT Chile to promote and protect human rights. For 2007, the project has planned various regional consultations with government actors and indigenous organizations concerning the state of implementation of the recommendations.

As in the case of Chile, the Special Rapporteur recommended to OHCHR that it provide technical cooperation to the Philippines for the promotion and protection of indigenous peoples' rights (see E/CN.4/2003/90/Add.3, para. 67 (j)). This recommendation, which has been endorsed and followed up by indigenous organizations, has not yet been implemented due to the lack of a technical cooperation project between OHCHR and the Government of the Philippines.

9.3.3 Follow-Up Initiatives by International Agencies

A number of international agencies have used the Special Rapporteur's thematic and country recommendations in their programmatic work. UNESCO, which took an active part in the preparation of the Special Rapporteur's thematic report on indigenous education,[3] has reportedly used the recommendations in that report in defining its general programs, particularly with regard to the promotion of bilingual education and the development of culturally appropriate curricula. The UNDP Regional Initiative on Strengthening Policy Dialogue on Indigenous, Highland and Tribal Peoples' Rights and Development (RIPP) has worked on access to justice, a question highlighted in the Special Rapporteur's second annual report, in Cambodia, the Philippines, Thailand and Viet Nam. UNHCR took note of the concern expressed by the Special Rapporteur regarding political violence against indigenous leaders in Colombia in the elaboration of its country assessment.[4]

---

[2] José Aylwin: "Implementación de las recomendaciones del informe de misión a Chile del Relator Especial de la ONU sobre los derechos humanos y las libertades fundamentales de los indígenas, Sr. Rodolfo Stavenhagen: experiencias y aprendizajes". Paper prepared for the International expert seminar on the implementation of the recommendations of the Special Rapporteur on the situation of human rights and fundamental freedoms of indigenous people (Montreal, 5–7 October 2006).

[3] Expert Seminar on Indigenous Peoples and Education: "Indigenous Education in the 21st century", organized jointly by OHCHR and UNESCO (Paris, 18–20 October 2004). The proceedings of the seminar are reproduced in document E/CN.4/2005/88/Add.4.

[4] UNHCR, International Protection Considerations Regarding Colombian Asylum-Seekers and Refugees (March 2005), para. 116.

In Guatemala, in keeping with the Special Rapporteur's recommendation, the Thematic Group on Indigenous and Multicultural Issues has continued operating as an inter-agency group of UNCT, involving indigenous peoples in its activities (see E/CN.4/2003/90/Add.2, para. 86). International agencies have further continued their cooperation in training indigenous peoples' organizations, a best practice that was also encouraged in the Special Rapporteur's report (ibid, para. 87). Similarly, various agencies of UNCT in Colombia are working together with the Kogui, Wiwa, Arhuaco and Kankuamo in the Sierra Nevada de Santa Marta region to elaborate a 'humanitarian diagnosis' of these peoples. This initiative aims at shedding light on their human rights situation taking into account their own perspectives and priorities.

Finally, the Special Rapporteur's reports have also informed the activities of the Inter-Agency Support Group providing technical assistance to the United Nations Permanent Forum on Indigenous Issues concerning the different issues covered at its annual sessions.

9.3.4 Follow-Up Initiatives by Civil Society

At the Montreal expert seminar indigenous leaders and experts concluded that they cannot wait for Governments to implement the recommendations of the Special Rapporteur. Rather, indigenous peoples and their support organizations, in cooperation with governmental and other non-governmental actors, should take a leading role in putting these recommendations into practice. A growing number of experiences in countries that the Special Rapporteur has visited provide examples of how indigenous peoples have appropriated these reports and used them as practical tools in the defense of their rights.[5]

A concern expressed by indigenous organizations in many of the countries visited by the Special Rapporteur is the lack of information among indigenous communities about his reports and recommendations. In order to address this shortfall, a number of indigenous organizations have promoted publication of the Special Rapporteur's reports. In the Philippines, Tebtebba published a book in 2002 which reproduced the Special Rapporteur's report on the country, as well as general information on the mandate. The book was widely disseminated nationally and abroad, and has helped indigenous peoples in other countries to make the best use of a mission by the Special Rapporteur. International NGOs working in the area of indigenous rights have focused on the activities undertaken by the Special

[5] Victoria Tauli-Corpuz and Erlyn Ruth Alcantara, Engaging the UN Special Rapporteur on Indigenous People: Opportunities and Challenges. The Philippine Mission of the UN Special Rapporteur on the Situation of Human Rights and Fundamental Freedoms of Indigenous People, Manila, Tebtebba-Indigenous Peoples' International Centre for Policy Research and Education, 2002.

Rapporteur.[6] Amnesty International (Canada) disseminated sections of the Special Rapporteur's report on major development projects as part of a national campaign to publicize the impacts of these projects on indigenous communities in the country. In Chile, the Lafkenche Mapuche published an abridged version of the Special Rapporteur's report and of the Chilean official response in 2005.

In Mexico, the Citizen Observatory of Indigenous Peoples (*Observatorio Ciudadano de los Pueblos Indígenas*, OCPI), established by the Mexican Academy of Human Rights, one of the main human rights NGOs in the country, in cooperation with the UNESCO Chair on Human Rights of the National Autonomous University of Mexico, monitors the implementation of the Special Rapporteur's recommendations after his visit to Mexico in 2003 to the States of Chiapas, Guerrero, Oaxaca, Puebla, Veracruz and Yucatán, the States with the highest density of indigenous populations in the country. The Observatory launched a nationwide campaign to promote knowledge of the Special Rapporteur's mandate and the recommendations of his report and evaluate the state of implementation of these recommendations through an information request system (SISI) about the different governmental programmes and projects aimed at the implementation of the recommendations, which is available to the general public via the Internet.[7]

Indigenous and civil society in a number of countries have also regularly promoted follow-up of the Special Rapporteur's recommendations though national consultations. In the Philippines, a national meeting, "Indigenous Peoples, the UN Declaration on the Rights of Indigenous Peoples and the Second Decade Programme of Action", was held in Manila in August 2005 and evaluated the state of implementation of the Special Rapporteur's recommendations following his visit to the country. A second meeting was held in February 2007, with the participation of the Special Rapporteur. A similar experience was the Open Forum, *Closing the Implementation Gap*, held in Ottawa in October 2006, organized by the *Assembly of First Nations* (AFN), the *Native Women's Association of Canada* (NWAC), the *Grand Council of the Crees* (Eeyou Istchee), Amnesty International (Canada) and the Canadian Friends Service Committee, which the Special Rapporteur attended.

Other relevant initiatives regarding the follow-up to the recommendations of the Special Rapporteur's country reports have been the organization of independent human rights observation missions to assess the state of implementation of these recommendations. An important initiative in this regard was the organization of the *International Mission of Verification on the Humanitarian and Human Rights Situation of Indigenous Peoples of Colombia* (IMV) in Colombia in October 2006. IMV was an initiative of the National Indigenous Organization of Colombia (*Organización Nacional Indígena de Colombia*, ONIC), in cooperation

[6] See e.g. "Bridging the Gap Between Law and Reality", in: *Cultural Survival Quarterly*, 30,1 (a special issue devoted to the seminar organized at the University of Arizona in cooperation with the Special Rapporteur in October 2005 on the implementation of domestic and international norms regarding the rights of indigenous peoples).

[7] See at: http://www.amdh.com.mx/ocpi.

with several indigenous and civil society organizations at the national and international levels. IMV visited the Sierra Nevada de Santa Marta and the Departments of Arauca, Cauca, Córdoba and Guaviare, and produced specific reports on the findings in those areas.

41. In other cases independent observation missions have focused on specific aspects of the Special Rapporteur's recommendations. In the case of Chile, Human Rights Watch and the International Federation of Human Rights conducted separate missions in 2004 and 2006, in cooperation with indigenous and civil society organizations, as a follow-up to the Special Rapporteur's recommendations concerning the criminal policy regarding Mapuche social protest in the south of the country, which in a number of cases has led to members of Mapuche communities receiving long prison sentences under the anti-terrorist legislation.

9.4 Best Practices in the Implementation of Recommendations

9.4.1 Canada

One of the most important developments that have taken place in recent years in Canada concerns reparations to victims of the Residential School system. Under this system several generations of Aboriginal children were compelled to attend schools far from their communities, leading to widespread psychological suffering, physical abuse and loss of identity. The system has been the object of an increasing number of court cases in recent years (see E/CN.4/2005/88/Add.3, paras. 60-61). The Special Rapporteur recommended that "special attention be paid to the nexus between the Residential Schools restitution process, the transgenerational loss of culture and its attendant social problems" (ibid, para. 102). This recommendation reportedly helped advance the negotiations towards the Indian Residential Schools Settlement Agreement, signed by the Government, the claimants, AFN and various Churches in May 2006. The agreement includes payments to former students who lived at one of these schools, a system to deal with serious claims of abuse, and an expedited system of compensation for the elderly. The agreement further funds programmes for healing, truth and reconciliation for former students and their families.

In the report on his visit to Canada the Special Rapporteur also paid specific attention to the high rates of violence experienced by indigenous women. Approximately 500 Aboriginal women have been murdered or reported missing over the past 15 years, and Aboriginal women are five times more likely to experience a violent death than other Canadian women (ibid, para. 56). In this connection, the Special Rapporteur recommended that "particular attention be paid by specialized institutions to the abuse and violence of Aboriginal women and girls, particularly in the urban environment" (ibid, para. 113). In March 2005, the

Government signed a five-year contribution agreement NWAC to run the 'Sisters in Spirit' program. This educational and policy program aims at addressing violence, particularly racialized and/or sexualized violence, against Aboriginal women through awareness-raising and practical-oriented research, aimed at gaining a better understanding of this phenomenon.

Another serious issue affecting indigenous women that was pointed out in the Special Rapporteur's report is the violation of property rights on Aboriginal reserves as a result of gaps in the existing legal regulation (ibid, para. 31). The Special Rapporteur called on the Government to address "with high priority the lack of legislative protection regarding on-reserve Matrimonial Real Property which places First Nation women living on reserves at a disadvantage" (ibid, para. 112). In June 2006, after a parliamentary committee published a report on the issue, the Government announced its intention to take legal steps to ensure legal protection of Aboriginal women's matrimonial real property. Since then, the Ministry of Indian Affairs, AFN and NWAC have led a process of consultation with representatives of over 630 First Nations to provide input for that proposal.

An important recent development is the reform of the Canadian Human Rights Act, whose section 67 exempts any actions taken by band councils and the Federal Government under the Indian Act from the application of the Act and from the system of petitions included in the Act. The Special Rapporteur specifically recommended that "the Canadian Human Rights Commission be enabled to receive complaints about human rights violations of First Nations, including grievances related to the Indian Act; and that section 67 of the Human Rights Act be repealed" (ibid, para. 108). In December 2006 the Government introduced legislation to repeal section 67, and when this reform enters into effect, indigenous peoples and individuals will have the ability to seek recourse with the Human Rights Commission. This measure is expected to increase the protection of indigenous peoples' rights, particularly those of Aboriginal women.

Despite these 'best practices' in the implementation of the Special Rapporteur's recommendations, participants in the Open Forum held in Ottawa in October 2006 expressed concern about the lack of institutional action in areas covered by these recommendations. A particularly controversial issue, also referred to by Members of Parliament in interviews with the Special Rapporteur, was Canada's negative vote on the United Nations Declaration of the Rights of Indigenous Peoples at the first session of the Human Rights Council, in March 2006. Efforts to reduce the gap in socio-economic indicators between indigenous peoples and the rest of Canadian society have been thwarted by the Government's failure to honor the Kelowna Accord, agreed to in November 2005 by the Federal Government, all the provinces and territories, and all the national Aboriginal organizations. Despite ongoing efforts to negotiate comprehensive land agreements, numerous conflicts still exist as a result of the failure to recognize indigenous property rights over indigenous lands, including the recent case of Caledonia, in Ontario.

9.4.2 Chile

After the Special Rapporteur visited Chile, the presidential Historical Truth and New Treatment Commission concluded its activities in 2003, and its final report coincides substantively with many of the Special Rapporteur's recommendations concerning the need for important reforms. One of these recommendations (see E/CN.4/2004/80/Add.3, para. 58) is the 'prompt ratification' of ILO Convention No. 169, as Chile is one of the few Latin American States that still have not ratified this fundamental instrument. The Government has taken substantive steps in this direction, and in June 2006, on the occasion of the National Day of Indigenous Peoples, formally expressed the commitment to "achieve, as soon as possible" the ratification of Convention No. 169. A recent international human rights observation mission assessed the state of the ratification process, which now depends on the support of only two senators.[8]

Positive signs have been reported concerning the change of the criminal policy towards the so-called 'Mapuche conflict' in the south of the country. The judicialization of the many existing conflicts over lands claimed by Mapuche communities in the south, and specifically, the application of the anti-terrorist legislation in a number of cases related to indigenous land claims, received particular attention in the Special Rapporteur's report on his 2003 visit. In this connection, the Special Rapporteur's report recommended not penalizing "legitimate protest activities or social demands by indigenous organizations and communities" and that the anti-terrorist legislation should not be applied in these cases (ibid, paras. 69–70).

Despite the Special Rapporteur's recommendations, judicial processes against Mapuche activities continued in recent years, leading to further long prison sentences. A new judicial process initiated in 2005 against members of Mapuche organizations, including some of those already serving prison sentences, for allegedly engaging in criminal 'illegal terrorist association', an accusation that became the object of a national and international outcry, prompted the Special Rapporteur to address an open letter to the President of Chile. The Court of Temuco eventually acquitted the defendants, and this acquittal marked a turning point in the judiciary's position concerning the unreasonable application of existing anti-terrorist legislation.

A hunger strike initiated in 2006 by the four convicts in the Poluco Pidenco case again brought domestic and international attention to this serious issue, and several mandate holders of the Human Rights Council addressed the Government in that regard. This led to a reconsideration of the criminal policy with regard to the land conflicts in southern Chile, and the recently elected President declared publicly that the anti-terrorist legislation would not be applied again in this context. The Government also introduced an initiative to reform the anti-terrorist

---

[8] FIDH, Misión de observación internacional. Chile: *Posibilidades de cambio en la política hacia los pueblos indígenas*, No. 456/3 (August 2006).

law, aimed at excluding from the scope of the crime of terrorism acts against property with no effect on the life and physical integrity of persons or the national security. The law is still pending consideration by the Senate.

The Special Rapporteur's recommendation to set up a program to reduce poverty among the country's indigenous communities (ibid, para. 62) has been the object of special consideration by the Government, notably the inclusion of the total indigenous population estimated to live in extreme poverty (73,500 people) under the system of social protection 'Chile in Solidarity' (*Chile Solidario*), launched in 2004. The Government has further continued implementing the program 'Origins' (*Orígenes*), an ambitious development project within the scope of the Indigenous Law (Law No. 19.253), with the support of the Inter-American Development Bank. Phase I of the project ended in 2006 with more than 3,000 projects implemented by the National Corporation on Indigenous Development (*Corporación Nacional de Desarrollo Indígena*, CONADI), and Phase II will be implemented in the period 2007-2011.

The above examples show that Chile has multiplied its efforts to improve the situation of indigenous peoples in recent years. However, these efforts are still thwarted by the limited recognition of indigenous peoples' rights in the existing legal and institutional framework. The constitutional reform adopted in November 2006 failed to include recognition of indigenous peoples and their rights, and subsequent proposals of constitutional reform fall very short of existing international standards and have not involved indigenous peoples. The Indigenous Land and Water Fund has proved an insufficient mechanism, partly due to the failure of the existing mechanism to affirm ancestral rights and to review irregular adjudication of indigenous lands in the past. Development projects continue to threaten the livelihood of indigenous communities in areas claimed as part of their traditional territories, as in the case of the Pascua Lama project in Atacama, opposed by the Diaguita community of Huasco Alto. Cases of police violence and abuse in indigenous communities have recently been documented, as in the case of the Temucuicui community. Meanwhile, the Mapuche convicted of terrorism continue to serve long prison sentences.

9.4.3 Colombia

The Special Rapporteur in the report on his visit to Colombia in 2004 expresses his concern about the threat of extinction hanging over 12 small groups of indigenous peoples living in the Amazon region who are experiencing a 'humanitarian emergency' as a result of armed conflict, illicit crops, environmental destruction and economic megaprojects (see E/CN.4/2005/88/Add.2, box, p. 16). Particularly worrisome is the situation of the Nukak Maku, an isolated hunter-gatherer community in the Department of Gavire. Their existence has become endangered in recent years as they have become embroiled in armed confrontations between guerrillas, paramilitaries and the Colombian Army, and as their lands have been encroached upon by coca growers. The number of community members that have been displaced from their traditional lands is now estimated at more than 200, approximately 50 per cent of the total population. The Special Rapporteur has addressed urgent appeals to the Government of Colombia on various occasions concerning the forced eviction of the Nukak and the killing of their leaders. The Special Rapporteur, together with the Special Adviser to the Secretary-General on the Prevention of Genocide, is currently involved in a dialogue with the Government concerning this pressing issue.

In June 2006, the Government presented a Plan for Integrated Assistance to Vulnerable Communities. The Plan includes special measures to attend to the urgent needs of the Nukak Maku, particularly in the fields of health and food security, as well as the temporary relocation of the displaced population in Puerto Ospina. This movement to areas that do not belong to the Nukak traditional territory has been the subject of controversy, and the recent suicide of a Nukak traditional leader has increased the international focus on the critical situation of this community. In a parallel initiative, OHCHR Colombia, in cooperation with the Office for the Coordination of Humanitarian Affairs, UNDP and UNHCR, has undertaken a comprehensive study on the situation of the Nukak Maku and have advised the Government on further possible actions to address it.

Another serious situation analyzed in the Special Rapporteur's report on Colombia is the selective killing and forced disappearance of indigenous leaders and traditional authorities, at the hands of both the guerrillas and the paramilitaries. By way of illustration, it offers the specific situation of the Embera-Katio people of Alto Sinú, who have suffered violence and intimidation because of their opposition to the construction of the Urrá hydroelectric dam on their territory, and who have been granted precautionary measures by the Inter-American Commission on Human Rights (ibid, box, p. 10). In connection with this and similar cases, the Special Rapporteur recommended that State authorities should immediately implement the precautionary measures granted by the Inter-American Commission to various indigenous communities. A positive development in this regard is the establishment of a mixed committee, comprised of government authorities, civil society, representatives of ONIC and authorities of the communities concerned, with OHCHR participating as an observer. The committee undertakes periodic visits to the region to verify the situation of the Embera-Katío and the state of implementation of the Commission's precautionary measures. The committee further requests specific government bodies to take action concerning the implementation of these measures.

In addition, in May 2005, the Government reached an agreement with the traditional authorities of the Embera–Katío to ameliorate the situation of the communities affected by the Urrá dam. The agreement consists of different measures in areas like the environment, education, health and food supply, including the elaboration of a plan to replace traditional hunting and gathering activities affected by the construction of the dam. The agreement further incorporates the Government's agreement to hold periodic meetings with indigenous representatives concerning the recommendations in the Special Rapporteur's reports. But much remains to be done to restore the livelihood of this endangered people.

Despite these specific cases in which the Government has taken action in favor of particularly vulnerable communities, the overall situation of indigenous peoples in Colombia has not improved since the Special Rapporteur visited the country. The International Verification Mission that visited several indigenous areas in 2006 concluded that indigenous people, and particularly women, are victims of serious human rights abuses and breaches of humanitarian law in the context of the ongoing armed conflict in the country, including selective killings, enforced disappearances, arbitrary detentions, torture and breaches of due process. Ongoing human rights violations against members of the Wiwa people and other communities of the Sierra Nevada de Santa Marta constitute a particularly serious example of this pattern. Indigenous organizations continue to denounce the impact of megaprojects on their traditional territories, as exemplified by the resumption of oil exploitation in the U'wa territory, in the Departments of Santander and Arauca, and the plans to construct a gas pipeline across the Wayuu traditional lands on the border with Venezuela.

9.4.4 Guatemala

The Special Rapporteur's recent follow-up visit to Guatemala allowed him to observe a number of changes and advances regarding the situation of indigenous peoples in the country in line with some of the recommendations included in the report on his 2002 visit. The Special Rapporteur noted in particular an increasing level of awareness among State authorities of the need to give priority attention to indigenous issues.

The Special Rapporteur's report on Guatemala paid special attention to the 1996 Peace Agreements, which include the Agreement on Identity and Rights of Indigenous Peoples. The agreement defines a comprehensive program of action to advance the recognition and protection of the rights of indigenous peoples (see E/CN.4/2003/90/Add.2, para. 4). Given the comprehensive character of these agreements, and the setback detected in their implementation, the Special Rapporteur recommended that the Government "carefully review the progress achieved in implementing the Peace Agreements insofar as they affect the indigenous peoples", and take "all appropriate measures to ensure full implementation" (ibid, para. 71). An encouraging development in this regard is the adoption in August 2005 of the Framework-Law on the Peace Agreement (Decree No. 52-2005), with the objective of regulating the implementation and monitoring of State action in this realm, and which makes the implementation of the Peace Agreements a legal commitment of the State.

In connection with the Peace Agreements, the Special Rapporteur also welcomed a number of initiatives to seek redress for the atrocities committed during the civil war. In 2004, in implementation of the decision of the Inter-American

Court of Human Rights in the Masacre de Plan de Sánchez case, concerning a massacre in a Mayan village in 1982 committed by the military, the Government organized a public event at which it acknowledged its responsibility for the atrocity and apologized to the victims and their relatives. The Presidential Commission on Human Rights (*Comisión Presidencial de Derechos Humanos*, COPREDEH) initiated in February 2006 a process of compensation of the victims of the massacre.

The Special Rapporteur's report emphasizes the need to strengthen and prioritize measures to combat the high level of racism and discrimination in the country. There have been a number of court decisions in recent years regarding cases of racial discrimination, which is a crime under the Guatemalan Penal Code. Institutional action in this regard has been reinforced with the establishment of the Presidential Commission to Combat Discrimination and Racism against Indigenous Peoples (*Comisión Presidencial contra la Discriminación y el Racismo contra los Pueblos Indígenas en Guatemala*, CODIRSA). As a follow-up to a specific recommendation in the Special Rapporteur's report (ibid, para. 67), CODIRSA, with the technical assistance of OHCHR Guatemala, has announced the launching in 2007 of a national campaign for coexistence and elimination of racism and racial discrimination.

Another issue of special concern that was pointed out in the Special Rapporteur's report on Guatemala is the situation of serious and systematic discrimination faced by indigenous women. In this regard, the Special Rapporteur recommended the adoption of 'special measures', including "greater political, legal and economic support to the *Office for the Defense of Indigenous Women* [*Defensoria de la Mujer Indígena*, DEMI]" (ibid, para. 79). A positive development in recent years has been the strengthening of the work of DEMI, with the support of international organizations and agencies, including OHCHR, UNDP, UNICEF and others. DEMI is now a key actor in the national human rights machinery, and requires continuous support to perform its important task.

The Special Rapporteur's report further recommends that Guatemala strengthen the educational system as a 'national priority', including the extension of bilingual education to all areas of the country (ibid, para. 77). An important measure of the implementation of this recommendation is the establishment of a Vice-Ministry of Bilingual Inter-cultural Education in 2003 and the adoption of Government Agreement No. 22-2004 on the extension of multicultural bilingual education in the education system, including the development of appropriate curricula. In addition, in 2003 Congress passed the Law on National Languages (Decree No. 19-2003), which officially recognizes the Mayan, Garifuna and Xinka languages and promotes their preservation and use in the Administration. This new legal and institutional framework has been welcomed by indigenous organizations and experts, who now demand its full implementation.

Despite these positive examples, and all the efforts deployed, the Special Rapporteur's second visit to Guatemala gave him the opportunity to ascertain that the levels of racism and discrimination against indigenous peoples are still worryingly high, and that the situation of indigenous women and children deserves urgent attention. The implementation of the Peace Agreements, and particularly of the *Agreement on Identity and Rights of Indigenous Peoples*, is thwarted by insufficient institutional backing and budgetary allocations. The justice system needs support to ensure that victims of human rights violations, and particularly indigenous women, find redress, and indigenous customary law needs to be recognized and incorporated in the work of the judiciary. Despite the acknowledgment of the atrocities committed in the past, the Special Rapporteur perceived that there will be no justice in Guatemala unless all those responsible for these acts are brought to justice.

9.4.5 Mexico

After a controversial constitutional reform was adopted in 2001, granting more powers to the states, many of the positive developments in the country concerning indigenous peoples' rights have taken place at the state level. Nevertheless, the federal constitutional review on indigenous issues remains at stalemate. State legislatures have followed the Special Rapporteur's recommendation to adopt legislation recognizing and protecting the rights of indigenous peoples (see E/CN.4/2004/80/Add.2, para. 66), including the Law on Indigenous Rights, Culture and Organization of Nayarit, Campeche and Quintana Roo.[9]

Important efforts have taken place to promote the implementation of the Special Rapporteur's recommendations concerning the review of the administration of justice in order to address indigenous peoples' specific needs (ibid, para. 82). Various initiatives have taken place to promote the consolidation and extension of the system of bilingual translators in courts, as recommended by the Special Rapporteur (ibid, para. 85). The Federal Government has undertaken a program of training of bilingual legal aid services, and in Oaxaca students at the Benito Juárez University work as bilingual legal aid lawyers. In Chiapas, the Office of the Prosecutor on Indigenous Justice (*Fiscalía de Justicia Indígena*) was created in 2005, and is staffed by indigenous lawyers who receive special training to ensure that the rights of indigenous peoples are respected in cases involving indigenous communities and individuals. In Querétaro, the Public Prosecutor's Office established a mobile office specializing in indigenous issues. Several states, including the States of México, Michoacán and Puebla, have started programmes to train legal translators and interpreters in indigenous languages.

In line with the Special Rapporteur's recommendation to incorporate indigenous law in the judicial system (ibid, para. 93), new 'indigenous courts' or 'peace and reconciliation courts' have been established in Campeche, Chiapas, Hidalgo, Puebla, Quintana Roo and San Luis Potosí, comprised of members of local

---

[9] CNDI, 2006: *La vigencia de los derechos indígenas en México*. Electronic book available at: http://cdi.gob.mx/derechos/vigencia_libro/vigencia_derechos_indigenas_mexico.pdf.

indigenous communities, with power to hear civil and family cases, as well as minor criminal cases, on the basis of indigenous law and custom. The National Commission for the Development of Indigenous Peoples (Comisión Nacional para el Desarrollo de los Pueblos Indígenas, CDI) has conducted studies on indigenous law and its 'compatibility' with human rights norms and national legislation.

The Special Rapporteur's recommendation to review the case files of indigenous persons prosecuted by the different courts in order to 'remedy any irregularities' (ibid, para. 86) 10 has been addressed by CDI, which has reviewed thousands of case files and is preparing a census of the indigenous population in national prisons. Similar programmes have been implemented in Hidalgo, Michoacán and Oaxaca.

A best practice is the implementation of the Special Rapporteur's recommendation to provide institutional strengthening of and adequate resources to bilingual intercultural education in the country (ibid, para. 102). The Ministry of Public Education has recently expanded bilingual secondary education, already provided in preschool and primary school, through a special course on indigenous peoples taught in several indigenous languages, and a number of 'intercultural high schools' and 'communitarian high schools', with adapted curricula and teaching in indigenous languages, have been created in areas of Chiapas, Oaxaca and Tabasco. Eight 'intercultural universities' have been set up in indigenous regions in the States of Chiapas, Guerrero, México, Michoacán, Puebla, Quintana Roo, Tabasco and Veracruz. The use of indigenous languages in education and in other spheres of public life has also been reinforced by the recently created National Institute on Indigenous Languages, responsible for the implementation of the General Law on the Linguistic Rights of Indigenous Peoples (2003).

Many of these best practices are the result of specific governmental and non-governmental initiatives to follow up on the recommendations of the Special Rapporteur (see paras. 21–23 and 38 above). Despite these positive steps, many important human rights concerns pointed out in the Special Rapporteur's recommendations have still not been addressed. The existing constitutional framework remains contested by many indigenous peoples and organizations and, notwithstanding the efforts of CDI, the reform has actually led to a lessening of the Federal Government's attention to indigenous issues. The agrarian legal and judicial system is obsolete in relation to the contemporary recognition of indigenous rights over their land and natural resources, and environmental policies have failed to sufficiently involve indigenous peoples, as in the case of the Montes Azules Biosphere Reserve. Development projects continue to threaten indigenous livelihoods, and the lack of clear consultation mechanisms has led to protracted conflicts, such as the case of the La Parota dam. The situation in Chiapas continues in a state of paralysis and human rights abuses by security forces and paramilitary groups have raised serious national and international concern, as exemplified by recent events in the State of Oaxaca.

The Special Rapporteur recommended particularly (para. 87) that CDI should be assigned a 'greater role' in this regard.

9.4.6 The Philippines

Information from different sources indicates that the Special Rapporteur's visit to the Philippines in 2003 has helped strengthen the country's institutional machinery with regard to the rights of indigenous peoples. The Special Rapporteur recommended, for instance, that the work of the *National Commission on Indigenous Peoples* (NCIP) should be supported "to become firmly established as the lead agency in protecting and promoting indigenous rights" with the widest possible participation of indigenous peoples (E/CN.4/2003/90/Add.3, para. 67 (a)). Since then, NCIP, with the support of international governmental and non-governmental donors, has strengthened its different lines of activity, particularly in relation to the delineation and recognition of *Certificates of Ancestral Domain Title* (CADTs) and the Ancestral Domain Sustainable Development and Protection Plan.

The Special Rapporteur's report further recommended that NCIP call for a "National Consultative Assembly" (ibid, para. 67 (a)), with the objective of including indigenous peoples and organizations in the planning and implementation of the Commission's activities. NCIP convened a National Forum in November 2006, leading to the establishment of the *Indigenous Peoples Consultative Body* (IPCB) operating at the national, regional and provincial levels. The composition of IPCB is tripartite, including representatives of NCIP, indigenous peoples' organizations and NGOs. Despite criticism concerning their membership, the establishment of these bodies has been seen as a positive development towards enhanced participation by indigenous peoples in the making and implementation of NCIP policies.

NCIP has strengthened its cooperation with the *National Commission on Human Rights* (NCHR) on indigenous issues. As recommended by the Special Rapporteur, NCHR has expanded its activities in the area of indigenous rights, including the development of training courses on the content of the Indigenous Peoples Rights Act for the police, the military, and other governmental bodies. Also in line with the Special Rapporteur's recommendation to promote special training programmes regarding the content of the Act (ibid, para. 67 (c)), the Government and civil society have concentrated efforts on training public officials, with special emphasis on members of the judiciary, with the cooperation of the Judicial Academy and the Ateneo Law School.

The Special Rapporteur's recommendations to extend education in indigenous areas (ibid, para. 67 (h)) and standardize the rights of indigenous peoples as at all levels of formal schooling (ibid, para. 67 (m)) were well received by the Department of Education, which in 2004 issued a permit to operate primary schools for indigenous peoples (Dep. Order No. 42). These schools can adapt their curriculum and calendar to the particularities of indigenous communities, and also incorporate 'para-teachers' from these communities in school teaching activities. Following the holding of the Third National Assembly on Indigenous Education in 2005, the Department of Education is currently embarked on a process of mainstreaming indigenous issues in the general curricula, in cooperation with professors of the University of the Philippines.

Significant advances have been reported in the implementation of the Special Rapporteur's recommendation to promote policy-oriented research by universities and civil society organizations regarding the rights of indigenous peoples (ibid, para. 67 (l)). National consultations were promoted in 2004 and 2005 by Tebtebba, the main indigenous research centre in the country, on strengthening the *Philippine Chapter of the Indigenous Peoples Global Research and Education Network*, an international network of individuals and institutions promoting indigenous research, education and development.

Nevertheless, the main areas of concern pointed out in the Special Rapporteur's report on the Philippines remain unaddressed. Despite the many efforts deployed by NCIP and its partners to promote the delineation and recognition of CADTs, NICP continues to be underfunded, and the rate at which titles are granted every year is still very limited in relation to the number of requests. Increased tension has been detected between the demarcation of indigenous lands and the agrarian reform promoted by the Department of Agrarian Reform, and certain indigenous territories have been identified as agrarian reform areas where individual titles are being granted to individual peasants. Serious human rights violations continue to be reported in relation to indigenous leaders and human rights defenders, a situation which was the subject of particular concern in the Special Rapporteur's report. Non-governmental sources have reported more than 75 cases of recent extrajudicial killings of indigenous individuals, many of which have not been thoroughly investigated.

9.5 Conclusions

The various cases reviewed in this study suggest that the Special Rapporteur's thematic and country reports have had a different level of impact. Inasmuch as they have the status of official United Nations documents elaborated from an independent viewpoint, thematic reports are part of ongoing discussions and policy-making concerning issues of special relevance for indigenous peoples, and their impact cannot be easily evaluated in terms of the implementation of the specific recommendations.

The Special Rapporteur's country visits have generally had a more direct impact on legal, social and political dynamics at the national level in relation to the recognition and protection of the rights of indigenous peoples. These reports, and the visits themselves, have helped promote spaces of dialogue between States and indigenous peoples; have contributed to educating government actors, civil society and the general public on the situation of indigenous peoples in their own countries; and have been appropriated by indigenous peoples and human rights organizations as an advocacy tool.

The recommendations included in the Special Rapporteur's reports do not provide a 'magic fix', and do not generate automatic and speedy changes in the situation of the rights of indigenous peoples. The level of implementation of these recommendations varies according to different country situations and the issues tackled by those recommendations.

Several initiatives have been undertaken over the last years by Governments, the United Nations system, civil society and indigenous organizations to monitor and promote the implementation of the recommendations included in the Special Rapporteur's reports. These experiences demonstrate that, if left for institutional action alone, the recommendations are rarely implemented, but implementation needs to be pushed forward in close cooperation with the Government and other stakeholders.

In countries where follow-up mechanisms exist, institutional efforts towards implementation have been more sustained, leading to concrete changes in law and practice.

These mechanisms have taken different forms, such as monitoring bodies, national forums and follow-up missions, and have involved a myriad of governmental and non-governmental actors, as well as international agencies.

The process of implementation of the Special Rapporteur's recommendations has opened spaces for dialogue between Governments, civil society and indigenous peoples and organizations. In all cases where substantive advances can be reported, indigenous peoples have been actively involved in the process.

The comparative analysis of best practices in several countries shows that the effective changes in implementation of the Special Rapporteur's recommendations are more easily detected in relation to recommendations related to the areas of social policy and development, as well as to the strengthening of specific government institutions and policies related to indigenous affairs. However, many of the main recommendations of the Special Rapporteur's reports remain unaddressed, particularly in the fields of legal and constitutional reform and indigenous land and resource rights, including the right of consultation in relation to development projects in indigenous territories.

These experiences suggest that, despite the advances that can be identified, the general record of implementation of the Special Rapporteur's recommendations is gloomy. Much remains to be done by the Governments, international agencies and other relevant stakeholders to bridge the 'implementation gap' that divides international and domestic norms and the serious human rights violations that indigenous peoples continue to experience in all parts of the world.

9.6 Recommendations

- Governments should multiply their efforts to promote effective changes in law and policy in implementation of the Special Rapporteur's recommendations, in compliance with international norms recognizing the rights of indigenous peoples.

- Governments should publicize and disseminate the Special Rapporteur's reports and recommendations among government institutions, civil society and indigenous peoples. Production of popular versions in various indigenous languages should be seriously considered.
- Governments should intensify their efforts to train public officials in the rights of indigenous peoples, taking into account the Special Rapporteur's reports and recommendations. The training of judges, prosecutors and public defenders based on these reports should be prioritized.
- The Governments concerned should establish permanent mechanisms to follow up on the recommendations of the Special Rapporteur's country reports. The mechanisms can include the designation of focal points to promote and coordinate efforts of different government departments and agencies such as interdepartmental working groups or specific units.
- Governments are encouraged to undertake periodic evaluations of the state of implementation of the Special Rapporteur's recommendations and to publicize the results.
- Governments should promote the involvement of indigenous peoples in the preparations for and carrying out of the Special Rapporteur's missions. Appropriate mechanisms should be put in place to promote the active participation of indigenous peoples in the implementation of the Special Rapporteur's recommendations.
- The Governments of Mexico and Guatemala are encouraged to continue the systematic follow-up to the recommendations initiated in close collaborations with OHCHR and indigenous peoples and organizations. The Governments of other countries that have been the object of an official visit by the Special Rapporteur are also encouraged to seek the technical assistance of OHCHR and international agencies in the implementation of the recommendations included in the reports on these visits.
- National parliaments, as well as national human rights institutions, are encouraged to take an active role in monitoring the implementation by all relevant actors of the Special Rapporteur's recommendations.
- Indigenous peoples and organizations, NGOs, academic institutions and other civil society actors are encouraged to strengthen their cooperation in order to foster the implementation of the Special Rapporteur's recommendations. They are also encouraged to use best practices from other countries concerning the establishment of permanent mechanisms and periodic initiatives to monitor the state of implementation.
- Indigenous peoples and their support organizations are encouraged to strengthen their involvement in the Special Rapporteur's general activities, including involvement in his country visits and dissemination of his reports.
- Public media are encouraged to pay increased attention to the Special Rapporteur's reports and visits, and to monitor the state of implementation of his recommendations.

- The Special Rapporteur invites OHCHR to incorporate, when applicable, the recommendations of his country and thematic reports in its program activities, particularly in relation to its field presence.
- OHCHR should continue its assistance to governmental institutions and civil society organizations to ensure follow-up to the Special Rapporteur's reports, taking into account the best practices described in this report.
- International organizations and agencies, including international financial institutions, should intensify their efforts to implement the Special Rapporteur's recommendations.
- United Nations country teams should designate a focal point to ensure the promotion and coordination of their activities in implementation of the Special Rapporteur's reports.
- International organizations and agencies should take into account the recommendations included in the Special Rapporteur's thematic reports in their programming in areas relevant to the rights of indigenous peoples. The Permanent Forum on Indigenous Issues Inter-Agency Group should also include these reports in the discussions on the topics analyzed at the Forum's annual sessions.
- International donors should support indigenous peoples and their support organizations to ensure their involvement in the Special Rapporteur's visits and other activities, as well as in their efforts to promote the implementation of his recommendations.

Chief Wilton Littlechild, a Creei ndigenous rights pioneer from Canada, addressing a meeting in Salekard, Russia. *Source:* Personal photographic collection of the author

About the Author

Rodolfo Stavenhagen (born 1932) is a Mexican anthropologist and sociologist. He is professor emeritus at El Colegio de México, one of Mexico's foremost social science institutions. From 2001 to 2008 he was United Nations Special Rapporteur for the Human Rights of Indigenous Peoples. Before that he was Assistant Director-General for Social Sciences at UNESCO, President of the Latin American Faculty of Social Sciences, and taught at numerous universities in Europe and the Americas. The University of Tromso, Norway, awarded him an Honorary Doctorate. He has worked on human rights, indigenous peoples, agrarian problems, social development and ethnic conflicts. Among his principal publications are *Social Classes in Agrarian Societies* (1975), *Ethnic Conflicts and the Nation-State* (1995), *The Ethnic Question: Development, Conflicts and Human Rights* (1990), *Derechos humanos y derecho indígena en América Latina (1989)*, and *Los pueblos indígens y sus derechos* (2008).

R. Stavenhagen, *Peasants, Culture and Indigenous Peoples*,
Texts and Protocols 4, DOI: 10.1007/978-3-642-34153-3, © The Author(s) 2013

About the Book

This third volume published on the occasion of the 80th birthday of Rodolfo Stavenhagen, professor emeritus of El Colegio de Mexico, includes eight essays on *Peasants, Culture and Indigenous Peoples: Critical Issues* that discuss: Basic Needs, Peasants and the Strategy for Rural Development (1976); Cultural Rights: a Social Science Perspective (1998); The Structure of Injustice: Poverty, Marginality, Exclusion and Human Rights (2000); What Kind of Yarn? From Color Line to Multicolored Hammock: Reflections on Racism and Public Policy (2001); The United Nations Special Rapporteur on the Rights of Indigenous Peoples (2012); A Report on the Human Rights Situation of Indigenous Peoples in Asia (2007); Report on the Impact of Megaprojects on the Rights of Indigenous Peoples (2003); Study Regarding the Best Practices to Implement the Recommendations of the Special Rapporteur (2007). These texts address human rights issues, especially those that arose when Stavenhagen was serving as United Nations special rapporteur on the rights of indigenous peoples.

R. Stavenhagen, *Peasants, Culture and Indigenous Peoples*,
Texts and Protocols 4, DOI: 10.1007/978-3-642-34153-3, © The Author(s) 2013

MIX
Papier aus verantwortungsvollen Quellen
Paper from responsible sources
FSC® C105338

If you have any concerns about our products,
you can contact us on
ProductSafety@springernature.com

In case Publisher is established outside the EU,
the EU authorized representative is:
Springer Nature Customer Service Center GmbH
Europaplatz 3, 69115 Heidelberg, Germany

Printed by Libri Plureos GmbH
in Hamburg, Germany